Biochemistry
The chemistry of life

David T. Plummer

Senior Lecturer in Biochemistry
King's College
University of London

330ᴸ

McGRAW-HILL BOOK COMPANY

London · New York · St Louis · San Francisco · Auckland · Bogotá
Guatemala · Hamburg · Lisbon · Madrid · Mexico · Montreal
New Delhi · Panama · Paris · San Juan · São Paulo · Singapore
Sydney · Tokyo · Toronto

Published by
McGRAW-HILL Book Company (UK) Limited
MAIDENHEAD · BERKSHIRE · ENGLAND

British Library Cataloguing in Publication Data
Plummer, David T. (David Thomas)
  Biochemistry—the chemistry of life.
  1. Biochemistry
  I. Title
  574.19'2
  ISBN 0-07-707208-1 ✓

Library of Congress Cataloging-in-Publication Data
Plummer, David T.
  Biochemistry, the chemistry of life / David Plummer.
     p.   cm.
  Includes index.
  ISBN 0-07-707208-1
  1. Biochemistry.   I. Title.
QP514.2.P57   1989
574.19'2—dc20   89-2622

1234 SP 8909

Typeset by Eta Services (Typesetters) Ltd, Beccles, Suffolk
Printed and bound in Great Britain by
Scotprint Ltd., Musselburgh

To Ruth
Malcolm and Sally
Jonathan and Suzanne
Martyn and Rebecca

John 1:3, 4 (A.V.) All things were made by him; and without him was not anything made that was made. In him was life; and the life was the light of men.

Contents

Preface

'Of making many books there is no end, and much study is a weariness of the flesh' (*Ecclesiastes* 12: 12). The words of Solomon must strike many a chord in the hearts of those involved in education and there are few, if any, who would disagree with the first part of this quotation. Each postal delivery brings more advertising material from publishers and when a new biochemistry book is published, one is entitled to ask 'why another one'?

Biochemistry: The Chemistry of Life is an introductory book, written for those students who find the size and detail of the large biochemistry texts rather daunting when they meet them for the first time. It aims to give readers an overall view of the essential features of the subject and is illustrated throughout with frequent examples of the applications of biochemistry to other disciplines.

Biochemistry: The Chemistry of Life is also aimed at students of the life sciences who need biochemistry as part of a degree, or professional qualification. Biochemistry is fundamental to an understanding of modern biology and all students need to study the subject as a major subsidiary to their main discipline. The book seeks to provide a comprehensive yet concise account of the subject with the emphasis on the relevance and application of biochemistry to the life sciences. It is suitable for many degree, professional and technical courses which require a working knowledge of biochemistry.

The book is also useful as background reading for teachers of A-level biology and chemistry.

I must, of course, thank all those colleagues at King's College who have helped me to produce this book. I am especially grateful to Dr Mike Perry for reading most of the manuscript and for his extremely valuable ideas, many of which I have adopted. I would also like to thank Dr Geoffrey Hall for his advice and suggestions. My thanks are also due to my former research students Dr David Obatomi, Dr Taiwo Fashola and Dr Mojgan Hossein-Nia for reading parts of the manuscript and to my colleague Dr Derek Evered who so gallantly volunteered to read the proofs.

Last but not least I must thank my wife Ruth for drawing some of the more difficult diagrams and helping me with the rest. Her love and encouragement, together with the patience of my family, have been very much appreciated during the preparation of the manuscript.

David Plummer
King's College, London

1. Introduction

1.1 What is biochemistry?

Biochemistry as the name implies (Gk *bios* = life) is the study of the chemistry of life and its roots lie in both chemistry and biology.

Chemistry

Biochemistry is concerned with the chemical nature and reactions taking place in living organisms so that a good foundation of chemistry is needed to understand the subject.

THE STRUCTURE OF BIOLOGICAL COMPOUNDS

Compounds in the biosphere can be considered from two points of view, both of which are relevant to biochemistry. The first of these is the static aspect which is mainly the study of *organic chemistry*. This looks at the chemical and physical properties of compounds found in living matter and an understanding of the chemical structures and properties of biomolecules is an important part of the subject. However, this knowledge alone is not enough since even if a complete chemical analysis of a living cell could be made, it would only give a picture of that cell at a particular point in time and the essential feature of life is its state of ceaseless change.

METABOLISM

The other area of biochemistry is dynamic and deals with the chemical and physical changes that take place in living organisms with time. This is known as *metabolism* and the chemical reactions in living organisms are extremely rapid. Many reactions take place in milliseconds (10^{-3} s), while movements within the molecules that give rise to the reactions take place in microseconds (10^{-6} s) and some even in nanoseconds (10^{-9} s).

Metabolism essentially consists of two types of pathways: those that involve the degradation and those concerned with the synthesis of complex molecules. The breakdown of molecules to their simpler constituents is known as *catabolism* and where this occurs energy is liberated. The opposite process, whereby molecules are assembled from simpler precursors, is known as *anabolism* and for this to occur energy has to be provided. Catabolism and anabolism are balanced in living matter which is in a state of dynamic equilibrium. However, the exact position of the steady state varies with time and depends on the growth of the organism and its physiological state.

Biology

THE STUDY OF LIVING MATTER

There are various levels of biological organization from the whole organism down to organs, tissues, cells, subcellular structures and eventually molecules (Fig. 1.1). Broadly speaking, *anatomy and physiology* look at living things from the point of view

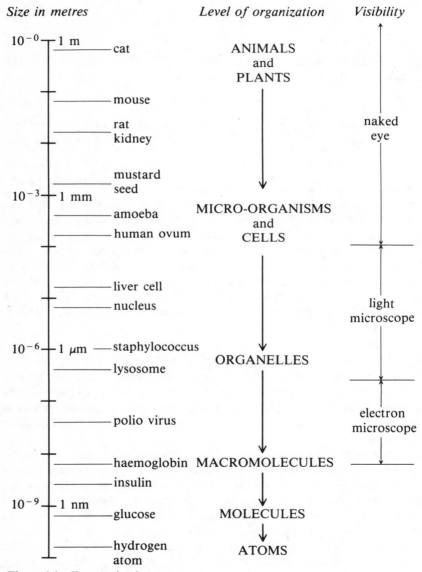

Figure 1.1 From animals to atoms.

of relatively large structures, such as organs and tissues, and seek to relate their function to that of the whole animal. *Biochemistry*, on the other hand, studies biological systems for the most part at the subcellular level and seeks to relate their structure and metabolism to that of the whole cell. The division between physiology and biochemistry is by no means rigid but, generally speaking, biochemistry is concerned with processes that occur at the subcellular and molecular level. Biochemistry therefore is the study of the molecular basis of life.

Figure 1.1 shows the relative sizes of the structures found in living organisms, together with the effective limits of the light and electron microscopes. The organizational divisions are by no means rigid and there is inevitably some overlap.

The basis of modern biology

Biochemistry is fundamental to modern biology and its study is now an essential part of all the life sciences. An understanding of the chemical events taking place at the molecular level can often give a useful insight into events involving the whole animal. For example neurochemistry, which is the biochemistry of the brain, has made a useful contribution to our understanding of animal and human behaviour. However, in the case of humans, if the whole basis of human personality and behaviour lies only in molecular processes then there is a distinct problem about responsibility and free will. Such questions are, of course, outside the realm of biochemistry and are really a matter for the philosopher and theologian to deal with. What is not in doubt is the valuable contribution made by biochemistry to many biological subjects including botany, zoology, microbiology, physiology, pharmacology, medicine and biotechnology.

The organization of the book

THE MOLECULAR BASIS OF LIFE

The first part of this book deals with the increasing complexity of the structures associated with living matter. It starts by looking at the elements and simple molecules of the biosphere, then moves on to consider the properties of some of the small organic molecules. This is followed by the very large macromolecules which are built up from these smaller units. Increasing structural complexity brings us to membranes and, finally, to the most complex structures of all, those of living cells and their organelles. A knowledge of the chemical and physical properties of these molecules and structures is extremely important and metabolism cannot be understood without mastering this biological chemistry and cell biology.

Part I concludes with a chapter on the properties of enzymes: those ubiquitous protein catalysts without which metabolism and life would be impossible.

THE CHEMICAL CHANGES IN LIVING MATTER

Part II of the text looks at metabolism and starts with thermodynamics and energy. Then follow details on how organisms obtain their energy by oxidation. Chapters 9 and 10 give examples of how this energy is used in the biosynthesis of important molecules and in detoxicating some natural and foreign metabolites. Chapter 11 then follows on molecular biology and some of its applications and the book concludes with examples to illustrate some of the principles of metabolic control.

Biochemistry is an important discipline in its own right but examples are given throughout the book to illustrate the importance and relevance of the subject to the biological and life sciences.

Part I The molecular basis of life

2. Simple molecules

2.1 Elements of the biosphere: the fundamental building blocks

There is an immense diversity of living organisms and each one contains a large number of organic and inorganic molecules. However, the elements that make up these molecules are relatively few and all of them have low atomic weights. In the case of the human body, 99.95 per cent of its total weight consists of only 11 elements and the remaining 0.05 per cent accounts for a further 13. Another four elements are known to have a function in plants and micro-organisms so only a small fraction of the 92 natural elements are found in living matter (Fig. 2.1). Furthermore, the frequency of occurrence of the elements in living systems is quite different to that found in non-living matter such as the atmosphere and rocks of our planet.

Major elements

The four most abundant elements in the biosphere are carbon, oxygen, hydrogen and nitrogen (C, O, H, N) and in the case of the human body they account for 96 per cent of the total wet weight. If the composition is expressed in terms of the number of

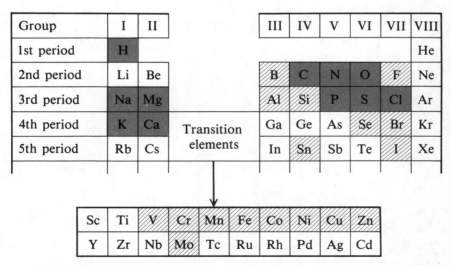

Figure 2.1 The elements of the biosphere. The first part of the periodic table showing the elements found in living matter. The major elements are shown in full colour and the minor elements are shaded.

atoms present, rather than the wet weight, then these four elements make up 99.4 per cent of the total. When looked at this way, 63 per cent of the atoms in the human body are hydrogen, the lightest and most insubstantial substance of all, so that as Shakespeare says 'We are such stuff as dreams are made on'.

Carbon (C) forms the basis of all organic molecules and an immense diversity of organic compounds is possible because of the ability of carbon atoms to form long chains by electron sharing. These chains may be linear, branched, cyclic or a combination of these so that very large structures (macromolecules) are possible.

The other major elements (O, H, N) also readily form strong covalent bonds with carbon and with each other and this increases still further the possible number, shape and size of the molecules that can be built up from these four elements. Fortunately for the study of biochemistry, the types of molecules that act as the building blocks of living matter are limited, although within each group there is quite a considerable variation.

Minor elements

The next seven most abundant elements account for only 4 per cent of the wet weight of the human body but they are vitally important (Fig. 2.2).

Calcium (Ca) and phosphorus (P) are the biggest constituents of the minor elements and in man most of the calcium and phosphorus is part of the structure of the

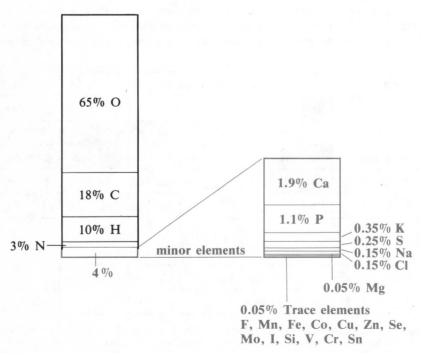

Figure 2.2 The elemental composition by weight of the human body.

bones. These elements are also important in metabolism as can be seen from the following examples. Calcium plays a key role in the regulation of cellular processes and also affects the contraction of heart and skeletal muscle. Phosphorus is an important part of the structure of the phospholipids, which form the basis of membranes, and the element is also present in many metabolic intermediates. It is a vital part of compounds involved in energy transfer such as adenosine triphosphate (ATP) which is often described as the 'energy currency' of the cell.

Sodium, potassium, magnesium, sulphur and chlorine (Na, K, Mg, S, Cl) are also important and adequate amounts of these elements must be supplied in the diet to maintain life. Sodium ions are the major cations of the extracellular fluid while potassium ions are predominant in the intracellular fluid. This ion differential is maintained by the constant expenditure of metabolic energy and is the driving force for membrane phenomena such as the initiation and transmission of the nerve impulse and membrane transport. This uneven distribution of ions between the extracellular and intracellular fluids is illustrated by the difference in cation composition of human blood plasma and liver cells (Fig. 2.3).

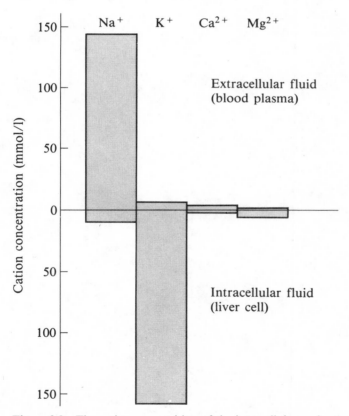

Figure 2.3   The cation composition of the intracellular and extracellular fluid in humans.

Trace and ultratrace elements

The remaining inorganic elements account for less than 0.05 per cent of the human body but nevertheless play a key role in maintaining life. For example, iron (<0.004 per cent) is needed for oxygen transport (haemoglobin) and oxidations (cytochromes) while iodine (<0.00004 per cent) is an essential part of the thyroid hormone that controls the metabolic rate. The role of trace and ultratrace elements in metabolism is an interesting area of current research and a deficiency or total lack of these elements in the diet can lead to quite serious diseases.

2.2 Water: the liquid environment of living organisms

Carbon, hydrogen, oxygen and nitrogen, the four most abundant elements in the biosphere, are present in the environment as simple molecules which are the starting point for the synthesis of more complex structures in living matter. Hydrogen and oxygen, for example, are present in water, a molecule which is absolutely essential for life. Water is not only the most abundant molecule on this planet but it is also the single biggest constituent of living organisms: the human body is about 70 per cent water and some plants contain 90 per cent or even more. The high concentration of water in living organisms means that many of the reactions in living matter take place in an aqueous environment. Life would be impossible without water, so it is important to understand the chemical and physical properties of this unique molecule.

Chemical structure

SIMPLE HYDRIDES

Water is chemically the hydride of oxygen and it has some unusual properties compared with the hydrides of sulphur, fluorine and nitrogen, elements that are immediately adjacent to oxygen in the periodic table (Fig. 2.1). Water, for example, has a much higher melting point and boiling point compared with the other hydrides which is perhaps surprising for such a small molecule.

	H_2O	H_2S	HF	NH_3
Melting point (°C)	0	−85	−83	−78
Boiling point (°C)	100	−61	20	−33

It is also a life-supporting liquid in contrast to hydrogen sulphide, hydrogen fluoride and ammonia which are all highly toxic.

MOLECULAR ASSOCIATION

The anomalous physical behaviour of water compared to the other hydrides is due to its molecular structure. The eight electrons shared between the atoms form a very stable configuration but the oxygen atom is strongly electronegative and attracts the bonding electron pairs of the O—H bonds away from the hydrogen atoms (Fig. 2.4). This gives the molecule a slight asymmetry of charge and a *hydrogen bond* is formed when the weak negative charge on the oxygen atom attracts the weak positive charge

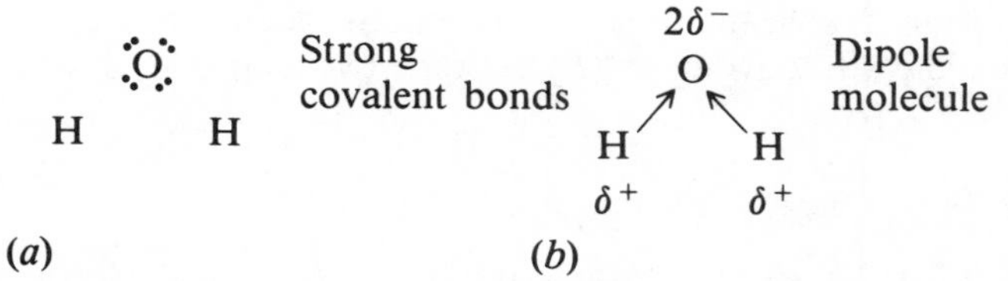

Strong covalent bonds   *Dipole molecule*

(a)   (b)

Figure 2.4 The molecular structure of water. (a) Stable shell of eight electrons. (b) Bonding electrons attracted towards the electronegative oxygen atom.

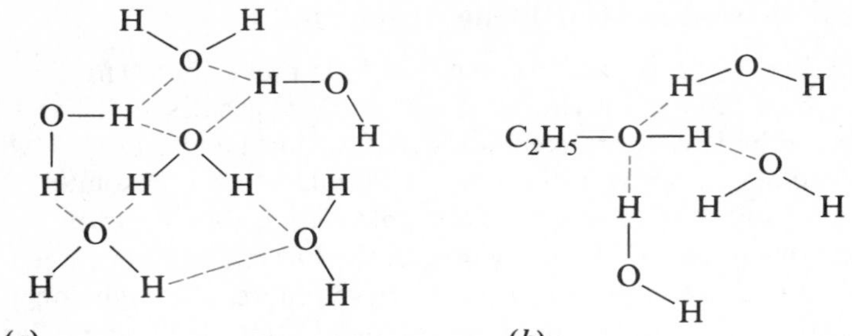

(a)   (b)

— strong covalent bonds

--- weak hydrogen bonds

Figure 2.5 The association of water molecules: (a) with other water molecules; (b) with ethanol, a polar molecule.

on the hydrogen atom of an adjacent water molecule. The force between the O and H atoms is weak but there are a large number of these hydrogen bonds and this gives a considerable internal cohesion to water (Fig. 2.5a). It also explains why water is such a stable liquid with a high melting point, boiling point and other unusual physical properties, many of which are essential to life (Table 2.1).

Table 2.1 The biological advantage of some of the physical properties of water

Physical property	Biological advantage
High specific heat	Absorption or radiation of heat can take place in living organisms with little change in temperature.
High surface tension	Capillary action is needed for the transpiration of water in plants.
High density at 4 °C	Ice forms on the surface of ponds and lakes at freezing point and aquatic life is preserved.
High heat of evaporation	Evaporation of water from the surface keeps animals and plants cool.

Physical properties

WATER AS A SOLVENT

Water dissolves most inorganic and organic compounds that ionize because of its high dielectric constant. The attractive force (F) between two univalent ions is given by:

$$F = e^+e^-/Dr^2$$

(e^+, e^- = charge, r = distance apart, D = dielectric constant)

In the solid state, D is small so that F, the attractive force between the ions, is large. However, in water, D is large so that F is much smaller and the ions separate and become dispersed throughout the liquid. The solubility of salts is further enhanced by the tendency of the resulting ions to become hydrated. Water is also a good solvent for organic compounds that do not ionize, provided they contain a sufficient number of polar groups. Solubility in this case is due to the association of the polar groups of the compound with water molecules by hydrogen bonds (Fig. 2.5b). Groups which tend to associate with water are said to be *hydrophilic* from two Greek words meaning 'water loving'.

HYDROPHOBIC INTERACTIONS

The strong tendency for water molecules to associate with each other is also the driving force for *hydrophobic interactions* which bring together non-polar molecules and groups. This is also derived from the Greek and is the exact opposite of hydrophilic and means 'water hating'.

If non-polar molecules are added to an aqueous medium, the water molecules that are in their immediate vicinity no longer move randomly and therefore become more ordered. However, the second law of thermodynamics states that systems tend to maximize their *entropy* or degree of disorder. To achieve this the non-polar molecules coalesce so as to give the minimum number of ordered water molecules and therefore the maximum entropy.

Some molecules contain both hydrophilic and hydrophobic regions and in an aqueous solution these *amphipathic molecules* can form ordered structures such as monolayers or micelles (Fig. 2.6). This is because the hydrophilic parts of the molecules seek to associate with the water and the hydrophobic regions with each other. Interactions similar to these are the driving force for the formation of phospholipid membranes and the 3-D structure of proteins: topics that will be discussed later.

IONIZATION

Water is very weakly ionized and the concentration of ions in pure water is extremely low.

$$H_2O \rightleftharpoons H^+ + OH^-$$

In practice the proton is hydrated as H_3O^+ so the equation for the dissociation of water is more strictly written as:

$$2\,H_2O \rightleftharpoons H_3O^+ + OH^-$$

Many biological processes are sensitive to H^+ and pH is a convenient way of describing the hydrogen-ion concentration in a solution.

$$pH = -\log_{10} [H^+]$$

At neutrality, $[H^+] = [OH^-] = 10^{-7}$ mol/litre so $pH = -\log_{10} [10^{-7}] = 7$ and the ionic product of water (K_w) is 10^{-14}.

$$K_w = [H^+][OH^-]$$

The concept of pH is very useful but it should always be remembered that the pH

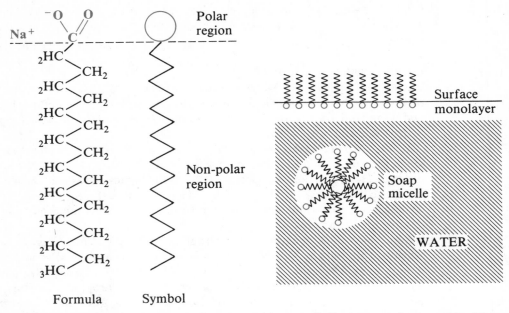

Figure 2.6 The association of sodium stearate with water. Sodium stearate is an amphipathic molecule which has a polar (hydrophilic) and a non-polar (hydrophobic) region. These can interact with water and each other to form a surface monolayer or a soap micelle as shown.

scale is logarithmic so that a change of one pH unit means an increase or decrease of ten-fold in the concentration of hydrogen ions [H^+]. The control of pH is very important since enzymes, metabolic processes and living cells can only function over a very narrow pH range. All cells therefore contain buffer systems to resist and control the changes in pH which would otherwise occur during metabolism. Just how narrow the limits of pH are can be seen in the case of human blood. This must be kept within a narrow range for life and an even narrower range for health. The mean blood pH is 7.4 and if this falls below 7.0 for any length of time, the person dies from acidotic coma. Similarly, if the pH remains above 7.8 then death occurs from tetany.

2.3 Gases of the atmosphere: interaction with the biosphere

The atmosphere of our planet plays a vitally important role in the biosphere as it is the source of the four main elements present in living matter. Hydrogen and oxygen are part of water vapour, carbon is a constituent of the gas carbon dioxide while nitrogen and oxygen are present as gaseous elements. Water and its importance to life have been discussed in Section 2.2 and this section gives an overall view of the involvement of carbon dioxide, oxygen and nitrogen in the living world.

The carbon cycle

The atmosphere contains only 0.03 per cent by volume of carbon dioxide but without this gas, life would not exist as it is the ultimate source of carbon, the basis of all

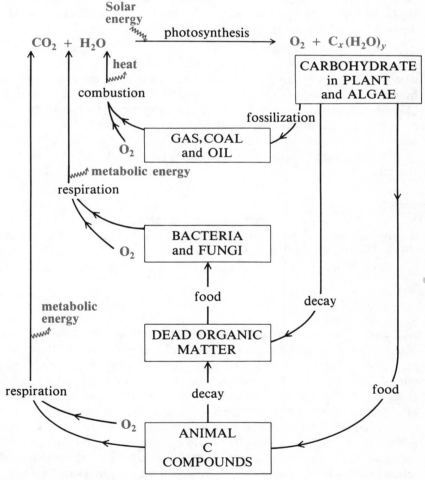

Figure 2.7 The carbon and oxygen cycle.

organic compounds. Oxygen, which makes up 21 per cent of the volume of air, is equally important since organisms oxidize carbon compounds to obtain the energy required to maintain life. These two molecules are involved in the transfer of carbon between the atmosphere and the biosphere in the carbon cycle which is illustrated in the diagram (Fig. 2.7).

INCORPORATION OF CARBON DIOXIDE INTO LIVING ORGANISMS

Carbon dioxide and water are incorporated into living matter during *photosynthesis* by green plants and algae. During this process the energy of sunlight is used to form complex organic molecules and liberate oxygen by the splitting of water. This is a complex process and the basic details are considered later in the book. The carbohydrates, which are formed by photosynthesis, are stored in plants and algae as food and the carbon passes along the food chain when the photosynthesizing organisms are eaten by land and aquatic animals.

RETURN OF CARBON DIOXIDE TO THE ATMOSPHERE

Almost all living organisms, with only a few exceptions, use oxygen to obtain energy by respiration. During this process the carbon atoms in the food are oxidized and liberated as carbon dioxide. When these organisms die they are subject to decay by micro-organisms which also oxidize the carbon compounds and release carbon dioxide. Sometimes the decaying organic matter ends up as fossil fuels (gas, oil, coal) so that the return of carbon to the atmosphere is delayed until fuels are burnt by combustion.

The nitrogen cycle

Nitrogen is the most abundant substance in the atmosphere and accounts for 78 per cent by volume of its total composition. It is another essential element as it is a vital part of the structure of proteins and nucleic acids without which life cannot exist. A summary of the transfer of nitrogen between the atmosphere and the biosphere is shown in Fig. 2.8.

NITROGEN FIXATION

Nitrogen is chemically inert and the $N \equiv N$ bond is very resistant to attack as it has a high bond energy of 940 kJ mol^{-1}. The reality of this is seen in the industrial Haber process which needs a high temperature (500 °C), a high pressure of several hundred atmospheres (3×10^7 Pa) and an iron catalyst in order to reduce nitrogen to ammonia. In view of this, the ability of nitrifying bacteria to convert atmospheric nitrogen into ammonia at atmospheric pressure and at low ambient temperature is quite remarkable. Other soil bacteria then convert the ammonia to nitrite and nitrate. It is estimated that more than 10^{11} kg of nitrogen are fixed by micro-organisms in a year and the details of this important process are considered later. Most of the nitrogen fixation takes place in the oceans and the soil but a small amount of nitrogen is oxidized to nitrate by the discharge of lightning during thunderstorms.

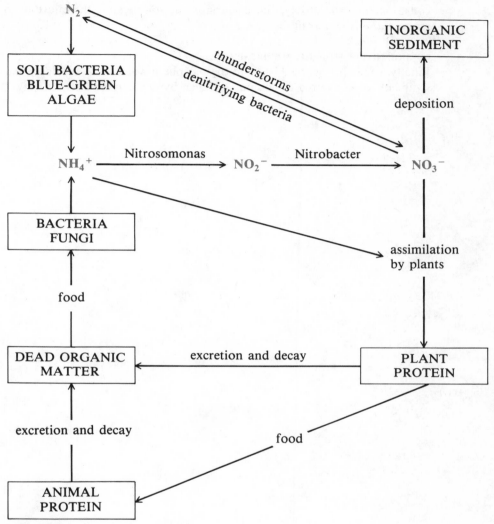

Figure 2.8 The nitrogen cycle.

TRANSFER OF NITROGEN

Nitrogen in the form of ammonia and nitrate is assimilated by plants and incorporated into proteins, nucleic acids and other nitrogen compounds. The nitrogen is then carried along the food chain when plants are eaten by animals on land and in the sea. During their lifetime, animals excrete the waste products of nitrogen metabolism as ammonia, uric acid or urea and these are then returned to the environment. When organisms die, their macromolecules are broken down to simpler constituents during decay, and bacteria in the soil degrade them further to produce ammonia, nitrite and nitrate.

LOSS OF NITROGEN FROM THE BIOSPHERE

Finally, some nitrogen is lost from the biosphere when nitrate is deposited in rocky sediments or converted to gaseous nitrogen by denitrifying bacteria.

3. Small organic molecules

3.1 Monosaccharides: the basic units of carbohydrates

Chemistry

MONOSACCHARIDES

Carbohydrates are carbon compounds that contain hydrogen and oxygen in the ratio of 2:1.

$$C_x(H_2O)_y$$

$$\uparrow \qquad \uparrow$$

carbon   hydrate

carbohydrate

This term is also used for derivatives of carbohydrates where the above definition may not be strictly true. Carbohydrates are of fundamental importance in living organisms where they are a source of metabolic energy and also form part of a number of important molecules and structures.

Hydrolysis of carbohydrates gives simple sugars or *monosaccharides* which are the building blocks of the larger carbohydrate molecules. Chemically they are polyhydroxyaldehydes (*aldoses*) or polyhydroxyketones (*ketoses*) and are named according to the number of carbon atoms that are present, so that *tetroses* contain four, *pentoses* five and *hexoses* six carbon atoms.

OPTICAL ISOMERS (D AND L SUGARS)

Sugars like many other biological molecules contain at least one *asymmetric carbon atom*. This is a carbon atom that has four different substituents and is also called a *chiral atom* or *chiral centre* (Fig. 3.1). The simplest aldose with one asymmetric carbon atom is glyceraldehyde and there are two possible ways of arranging the four different groups around the chiral centre. The structures of these two isomers are shown in Fig. 3.1 where the asymmetric carbon is at the centre of a tetrahedron and the four groups at each of the corners. These two forms are non-superimposable mirror images and can also be likened to left- and right-handed gloves. The two configurations are *optical isomers* or *enantiomers* and rotate the plane of polarized light to the same extent but in opposite directions. The *dextrorotatory* form rotates the plane of polarized light to the right and is designated (d) or (+), while the *laevorotatory* compound rotates the plane of polarized light to the left and is identified as (l) or (−).

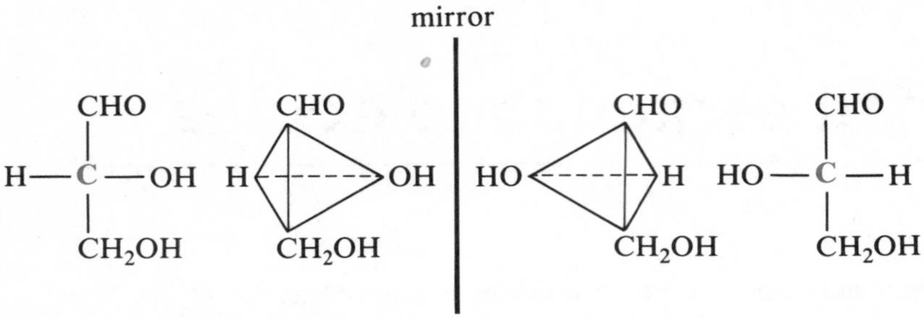

mirror

$$
\begin{array}{ccc}
\text{CHO} & \text{CHO} & \\
| & & \\
\text{H—C—OH} & \text{H} & \text{OH} \\
| & & \\
\text{CH}_2\text{OH} & \text{CH}_2\text{OH} &
\end{array}
\qquad
\begin{array}{ccc}
\text{CHO} & \text{CHO} \\
& | \\
\text{HO} \quad \text{H} & \text{HO—C—H} \\
& | \\
\text{CH}_2\text{OH} & \text{CH}_2\text{OH}
\end{array}
$$

D(+)-glyceraldehyde parent
compound for D-aldoses

L(−)-glyceraldehyde parent
compound for L-aldoses

(Asymmetric carbon atom C)

Figure 3.1   Optical and stereoisomers.

STEREOISOMERS (D AND L SUGARS)

The addition of successive secondary alcohol groups (—CHOH) to the parent compound, glyceraldehyde, gives rise to a whole family of aldoses. Since each new secondary alcohol group is a chiral centre, the number of possible *stereoisomers* rises rapidly with increasing number of carbon atoms. Aldoses that are derived from d-glyceraldehyde are known as D sugars and those arising from l-glyceraldehyde are called L sugars. The letters D and L refer to the spatial arrangement of atoms round the asymmetric carbon atom most distant from the carbonyl group and not to the direction of rotation of polarized light. Thus, a D sugar could be dextrorotatory [D(+)] or laevorotatory [L(−)] depending on the configuration of the other carbon atoms present.

The simplest ketose is the triose, dihydroxyacetone, but this compound is optically inactive so the parent compound of the ketoses is erythrulose which is a tetrose.

$$
\begin{array}{l}
\text{CH}_2\text{OH} \\
| \\
\text{C=O} \\
| \\
\text{H—C—OH} \\
| \\
\text{CH}_2\text{OH}
\end{array}
$$

D-erythrulose

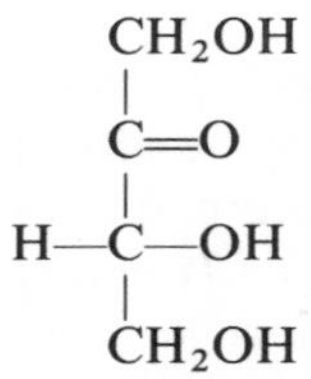

The majority of sugars present in living matter are of the the D configuration although there are some exceptions to this in micro-organisms.

RING FORMULAE (α AND β FORMS)

If D(+)-glucose is dissolved in water the rotation of the plane of polarized light slowly falls with time, before stabilizing at a new value some 24 h later. This phenomenon is known as mutarotation and is shown by a number of sugars. The reason for the

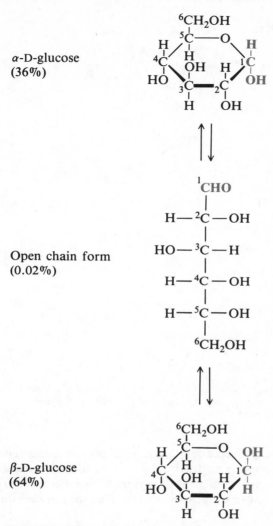

Figure 3.2 Forms of D-glucose (per cent at equilibrium). The carbon atoms are numbered from the end of the chain that contains the reactive aldehyde or ketone group.

change in rotation is that glucose exists in solution mainly in the ring form, which is an intramolecular hemiacetal (Fig. 3.2). This creates another asymmetric carbon atom (C-1) so that two ring forms (α and β) are now possible and these are known as *anomers*.

The formulae shown in Figs. 3.2 and 3.3 are those proposed by Haworth and show the ring as a plane with the hydroxyl groups orientated above and below the plane. For the sake of clarity the hydrogen atoms are not usually shown.

HOH$_2$C   O   OH

HO   CH$_2$OH

OH

(a) β-D-fructofuranose (a hexose)

CH$_2$OH

HO   O

OH

OH

OH

(b) α-D-galactopyranose (a hexose)

HOH$_2$C   O   OH

OH

(c) β-2-deoxy-D-ribose (a pentose)

Figure 3.3 Haworth formulae of sugars. The ring is shown as a plane with the bold lines nearest the reader. The functional groups then lie above or below this plane.

PYRANOSE AND FURANOSE

The ring formed in the case of glucose contains five carbon atoms and one oxygen and is analogous to pyran; glucose is said therefore to exist in the *pyranose* ring form. Many of the common sugars are present as the pyranose ring, but some exist as a ring containing four carbons and one oxygen and this is called the *furanose* form after the compound furan.

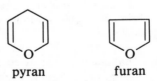

pyran   furan

Biological function

FUEL FOR OXIDATION

A number of sugars are metabolic fuels which can be broken down and oxidized to produce energy.

Glucose   This is one of the major fuels of the cell and in animals ingested starch is broken down to glucose. This is then transported in the blood to the tissues where it is oxidized to carbon dioxide and water during cellular respiration. The energy released during this oxidation is then 'captured' and used to drive the metabolic machinery of the cells. If there is an excess of glucose, this is then stored until needed as a polymer in the form of starch in plants and as glycogen in animals.

Fructose   This is the sweetest of all the sugars and is a store of food in fruit and honey. Fructose is found in high concentrations in human semen where it is oxidized to provide the energy for the migration of the spermatozoa in the female tract.

Lactose   This compound is a disaccharide of galactose and glucose (β-D-galactosyl-β-1,4-D-glucose) and is the sugar present in mammalian milk. During digestion this is broken down in the gut to its constituent monosaccharides which are then absorbed and oxidized.

Sucrose   This is another disaccharide and consists of glucose and fructose joined through both of their potential reducing groups (α-D-glucosyl-β-1,2-D-fructose). Sucrose is a food store in some plants and high concentrations are found in sugar cane and sugar beet. The sugar is a major source of carbohydrate in developed countries and during digestion it is broken down to its constituent monosaccharides which are then absorbed and oxidized.

SIMPLE MONOSACCHARIDE DERIVATIVES

Derivatives of monosaccharides are also found in nature and some of the more important compounds are mentioned below.

Oxidized products   The aldehyde and primary alcohol groups can be oxidized to the corresponding aldonic and uronic acids. The oxidation of a monosaccharide is shown

$$
\begin{array}{ccc}
 & \text{CHO} & \\
 & | & \\
 & (\text{CHOH})_n & \\
 & | & \\
 & \text{CH}_2\text{OH} & \\
\text{Pt/O}_2 \swarrow & \text{monosaccharide} & \searrow \text{Br/H}_2\text{O} \\
 & \text{(glucose)} & \\
\text{CHO} & & \text{COOH} \\
| & & | \\
(\text{CHOH})_n & & (\text{CHOH})_n \\
| & & | \\
\text{COOH} & & \text{CH}_2\text{OH} \\
\text{uronic acid} & & \text{aldonic acid} \\
\text{(glucuronic acid)} & & \text{(gluconic acid)}
\end{array}
$$

above and the names in brackets represent the products obtained when the sugar is glucose. Derivatives of these compounds feature in carbohydrate metabolism and conjugation with glucuronic acid is important in the excretion and elimination of steroids and drugs from the body.

Reduction The carbonyl group present can be reduced to the primary or secondary alcohol with sodium borohydride. For example, glucose and fructose can be reduced to the hexahydric alcohol *sorbitol*, which is used as an artificial sweetener, and glyceraldehyde may be reduced to *glycerol*, the trihydric alcohol which forms fat when esterified with long-chain fatty acids.

$$
\begin{array}{ccc}
\text{CHO} & & \text{CH}_2\text{OH} \\
| & & | \\
\text{CHOH} & \xrightarrow{\text{[2H]}} & \text{CHOH} \\
| & & | \\
\text{CH}_2\text{OH} & & \text{CH}_2\text{OH} \\
\text{glyceraldehyde} & & \text{glycerol}
\end{array}
$$

Esters Sugars form esters very readily with acids and the phosphate esters of a number of sugars are important intermediates in carbohydrate metabolism. Esters formed from the primary alcohol group of the ultimate carbon (glucose-6-phosphate) are generally more stable than hemiacetals formed through the reducing group (glucose-1-phosphate).

Amino sugars The hydroxyl group in the 2-position of many sugars can be replaced by an amino group, as for example in *galactosamine* and *glucosamine*. These compounds also exist as the *N*-acetyl derivatives which are part of the structure of several polysaccharides, including chitin of the exoskeleton of insects and crustaceans such as crabs and lobsters.

3.2   Amino acids: the building blocks of proteins

Chemistry

GENERAL STRUCTURE

The breakdown of proteins by hydrolysis gives rise to amino acids which can be regarded as the building blocks of these important macromolecules. As the name implies, amino acids contain basic amino and acidic carboxyl groups and are therefore *amphoteric molecules* having both basic and acidic properties. The amino acids commonly found in proteins have the amino and carboxyl groups attached to the same carbon atom and are therefore all α-amino acids.

$$
\begin{array}{ccc}
 & COOH & COO^- \\
 & | & | \\
An\ \alpha\text{-amino acid} \quad & {}_2HN-C-H & {}_3HN^+-C-H \\
 & | & | \\
 & R & R \\
 & \text{un-ionized form} & \text{zwitterion}
\end{array}
$$

The chemical properties of the amino acids are greatly affected by the nature of the R group and the structures of the amino acids commonly found in proteins are given in Table 3.1.

STEREOCHEMISTRY

The α-carbon atom is asymmetric and is a chiral centre (**C**) for all amino acids except glycine so that, apart from glycine, all amino acids show optical activity. If L-serine is taken as a typical amino acid, then this can be compared with L(−)-glyceric acid, which is derived from L(−)-glyceraldehyde, the parent compound for the L series of sugars. When —CH$_2$OH is replaced by other groups, two families of amino acids emerge, the L and the D series. All of the amino acids found in proteins are of the L configuration although D amino acids are found in antibiotics and bacterial cell walls. Some amino acids, such as isoleucine and threonine, contain a second chiral centre so that more than two forms are possible.

$$
\begin{array}{ccc}
 & & \text{mirror} \\
COOH & COOH & | \quad COOH \\
| & | & | \quad | \\
HO-C-H & {}_2HN-C-H & | \ H-C-NH_2 \\
| & | & | \quad | \\
CH_2OH & CH_2OH & | \quad CH_2OH \\
L(-)\text{-glyceric acid} & L(+)\text{-serine} & \quad D(-)\text{-serine}
\end{array}
$$

This difference in the spatial configuration of the groups of the L and D amino acids is small but can be extremely important. For example, the antibiotic penicillin prevents the incorporation of the D amino acids into the cell walls of some bacteria but has no effect on the metabolism of L amino acids in animal cells.

Table 3.1 The common amino acids found in proteins
All of the amino acids are α-amino acids of the L configuration with the exception of proline and hydroxyproline which are imino acids

Name	Symbol	Structure
Aliphatic amino acids		
Glycine	Gly	$H—CH(NH_3{}^+)COO^-$
Alanine	Ala	$CH_3—CH(NH_3{}^+)COO^-$
Valine	Val	$CH_3\!\!\diagdown$ $CH(NH_3{}^+)COO^-$ $CH_3\!\!\diagup$
Leucine	Leu	$CH_3\!\!\diagdown$ $CH—CH_2—CH(NH_3{}^+)COO^-$ $CH_3\!\!\diagup$
Isoleucine	Ile	$C_2H_5\!\!\diagdown$ $CH_2—CH(NH_3{}^+)COO^-$ $CH_3\!\!\diagup$
Hydroxyl amino acids		
Serine	Ser	$HO—CH_2—CH(NH_3{}^+)COO^-$
Threonine	Thr	$HO—CH—CH(NH_3{}^+)COO^-$ $\mid$ CH_3
Sulphur amino acids		
Cysteine*	Cys	$HS—CH_2—CH(NH_3{}^+)COO^-$
Methionine	Met	$CH_3—S—CH_2—CH_3—CH(NH_3{}^+)COO^-$
Acidic amino acids and their amides		
Aspartic acid	Asp	$^-OOC—CH_2—CH(NH_3{}^+)COO^-$
Asparagine	Asn	$_2HNOC—CH_2—CH(NH_3{}^+)COO^-$
Glutamic acid	Glu	$^-OOC—CH_2—CH_2—CH(NH_3{}^+)COO^-$
Glutamine	Gln	$_2HNOC—CH_2—CH_2—CH(NH_3{}^+)COO^-$

Name	Symbol	Structure

Basic amino acids

Lysine   Lys $_3\overset{+}{H}N-(CH_2)_4-CH(NH_3{}^+)COO^-$

Arginine   Arg $\begin{array}{c} _2HN \\ \\ _2\overset{+}{H}N \end{array}\!\!\!>\!\!C-NH-(CH_2)_3-CH(NH_3{}^+)COO^-$

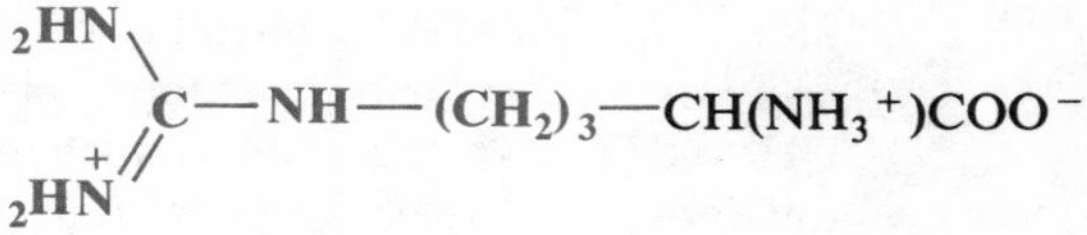

Aromatic and heterocyclic amino acids

Phenylalanine   Phe $\langle\!\!\bigcirc\!\!\rangle-CH_2-CH(NH_3{}^+)COO^-$

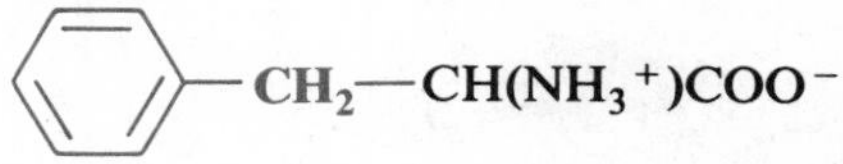

Tyrosine   Tyr $HO-\langle\!\!\bigcirc\!\!\rangle-CH_2-CH(NH_3{}^+)COO^-$

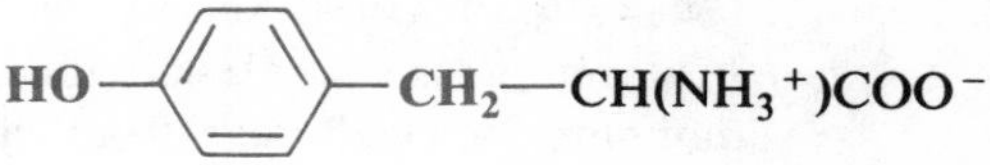

Tryptophan   Trp   (indole ring)$-CH_2-CH(NH_3{}^+)COO^-$

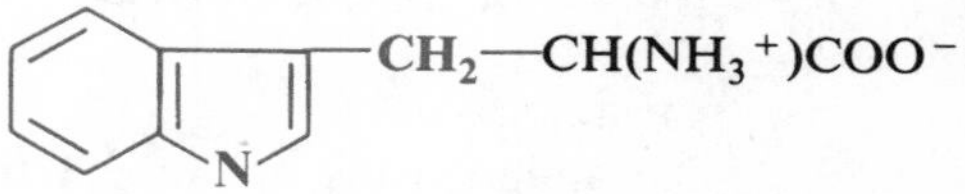

Histidine   His   (imidazole ring, HN⋯N)$-CH_2-CH(NH_3{}^+)COO^-$

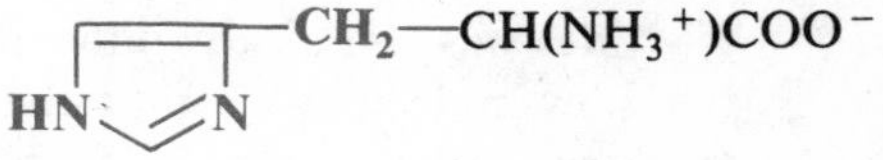

Imino acids

Proline†   Pro   (pyrrolidine ring, $\overset{+}{N}H_2$)$-COO^-$

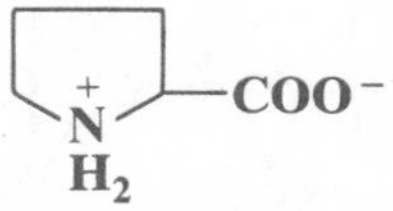

*Two cysteine residues linked by a disulphide bond form the amino acid cystine (Cys — Cys).

†Hydroxylation of proline in the protein molecule gives the imino acid hydroxyproline (Hyp).

IONIZATION

Amino acids have unexpectedly high melting points for such small molecules; for example glycine ($_2$HN—CH$_2$—COOH) is a crystalline solid which melts at about 230 °C, whereas the structurally related compound methylamine ($_2$HN—CH$_3$) is a gas and acetic acid (CH$_3$—COOH) is a liquid at room temperature. The reason for the high melting points is the strong electrostatic attraction between the charged atoms which makes it difficult to disrupt the crystalline lattice. This is because glycine and other amino acids exist mostly in the doubly ionized or *zwitterion* form ($^+_3$HN—CH$_2$—COO$^-$) rather than as uncharged molecules. Amino acids are therefore unlike low-molecular-weight organic compounds in their properties and resemble inorganic salts.

The side chains (R) of the amino acids can also be ionized so that molecules may carry a net negative or positive charge, depending on the pH of the medium, and these ionizable groups are the basis of the charge properties of peptides and proteins (Table 3.2).

Function in the living organism

FORMULAE OF AMINO ACIDS IN PROTEINS

Just as monosaccharides are the basic units of polysaccharides, so amino acids can be thought of as the 'bricks' from which the protein 'house' is built. Polysaccharides are built up of only a few monosaccharide units, but proteins may contain as many as 22 different amino acids. The names and abbreviations of the amino acids that are found in proteins are given in Table 3.1.

PEPTIDE BOND

The amino acids are joined together in the protein molecule by peptide bonds (—CO—NH—) formed by the condensation of the α-carboxyl and the α-amino groups of different amino acids with the elimination of water. The chemical synthesis of peptides was technically difficult for many years but can now be achieved automatically. Low-molecular-weight polymers of amino acids are known as polypeptides while the term protein is reserved for larger polymers with molecular weights of several thousand or more.

$$_3\text{HN}^+\!\!-\!\text{CH}\!-\!\text{COO}^- \qquad + \qquad _3\text{HN}^+\!\!-\!\text{CH}_2\!-\!\text{COO}^-$$
$$| \phantom{_3\text{HN}^+\!\!-\!\text{CH}\!-}$$
$$\text{SH}$$

$$\downarrow$$

$$_3\text{HN}^+\!\!-\!\text{CH}\!-\!\text{CO}\!-\!\text{NH}\!-\!\text{CH}_2\!-\!\text{COO}^- \quad + \quad \text{H}_2\text{O}$$
$$|$$
$$\text{SH}$$

cysteinylglycine (a dipeptide)

Table 3.2 Ionizable groups of some amino acids

Amino acid	Ionizing group	pK_a	Charge at pH 7
Aspartic acid	β-carboxyl	3.9	99.9% −
Glutamic acid	γ-carboxyl	4.2	99.8% −
	$-COOH \rightleftarrows -COO^- + H^+$		
Histidine	imidazole	6.1	11.2% +

Cysteine	sulphydryl	8.0	9.1% −
	$-CH_2SH \rightleftarrows -CH_2S^- + H^+$		
Tyrosine	phenolic	10.1	0.1% −

Lysine	ε-amino	10.5	100 % +
	$-NH_3^+ \rightleftarrows NH_2 + H^+$		
Arginine	guanidine	12.5	100 % +

AMINO ACIDS AND PEPTIDES IN BIOLOGY

Free amino acids are present as a result of the normal turnover of proteins and some of them are metabolized to highly active molecules which play a key role in the living organism. Similarly, many small peptides show considerable biological activity and act as hormones, opiates or antibiotics. A few examples of these physiologically and pharmacologically active molecules are given in Fig. 3.4.

Peptide chains are also found attached to carbohydrate material to form the peptidoglycans of bacterial cell walls and the glycoprotein of blood group substances. In the former case many of the amino acids are of the D and not the usual L configuration.

Amino acid derivatives

Asparagine

COO^-
|
$CHNH_3^+$
|
CH_2CONH_2

Nitrogen
store
in plants

Adrenaline

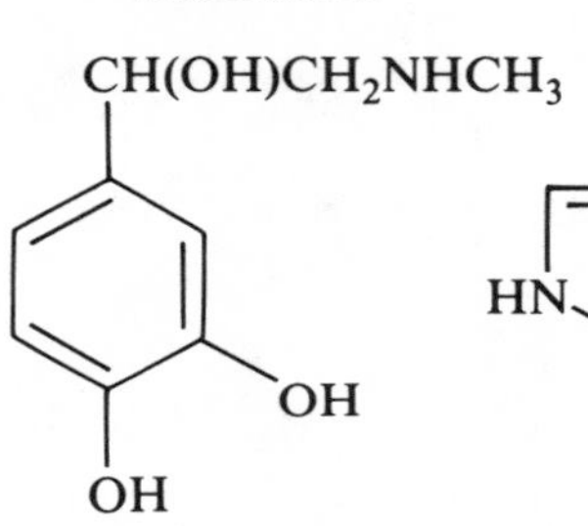

$CH(OH)CH_2NHCH_3$

Elevates
blood
pressure

Histamine

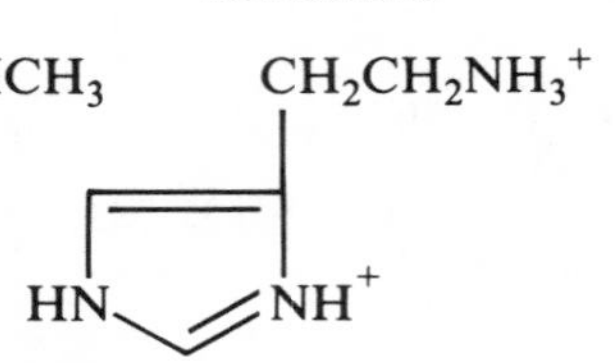

$CH_2CH_2NH_3^+$

HN NH^+

Liberated in
shock and
inflammation

γ-Aminobutyrate

COO^-
|
$(CH_2)_3$
|
NH_3^+

An
inhibitory
neurotransmitter

Peptides

Vasopressin

Cys—Tyr—Phe—Gln—Asn—Cys—Pro—Arg—Gly—NH$_2$

A hormone which acts on the kidney to retain water in the body.

Leucine enkaphalin

Tyr—Gly—Gly—Phe—Leu

A natural opiate present in the brain with strong pain-killing properties

Gramicidin

D-Phe—L-Leu—L-Orn—L-Val—L-Pro
| |
L-Pro—L-Val—L-Orn—L-Leu—D-Phe

An antibiotic from the bacterium *Bacillus brevis* with the unusual amino acids D-phenylalanine and L-ornithine.

Figure 3.4   Some examples of amino acid derivatives and peptides of biological importance.

3.3 Triacylglycerols: the long-term energy stores

Chemical and physical properties

LIPIDS

The term lipids is used to describe fatty substances that are insoluble in water but soluble in fat solvents such as ethanol, chloroform and ether. Chemically they are esters of long-chain fatty acids and are classified according to the nature of the alcohol moiety. Alkaline hydrolysis of lipids gives an alcohol and the sodium or potassium salts of the constituent fatty acids. This is known as *saponification* and the products of hydrolysis are often water soluble unlike the original lipids. Chemically, lipids can be divided into two main groups: *simple lipids* and *compound lipids*. Steroids and the fat-soluble vitamins are also classified as lipids because of their solubility characteristics and these compounds are known as *derived lipids*. However, many of the derived lipids are alcohols and not esters so that they cannot be saponified.

FATTY ACIDS

All lipids contain fatty acids which are organic acids with a single carboxyl group at the end of a long hydrocarbon chain. A wide variety of fatty acids are found in lipids but most of them have straight chains with an even number of carbon atoms. Natural fatty acids may contain anything from 4 to 24 carbons but the most abundant are those with 16 or 18 carbon atoms (Table 3.3). Fatty acids may be fully saturated but the majority of natural compounds are unsaturated molecules containing one or more double bonds (Table 3.3).

Table 3.3 Some common fatty acids

Common name	Formula	Systematic name	Symbol*	MP(°C)
Saturated fatty acids				
Lauric	$CH_3(CH_2)_{10}COOH$	Dodecanoic	12:0	44
Myristic	$CH_3(CH_2)_{12}COOH$	Tetradecanoic	14:0	54
Palmitic	$CH_3(CH_2)_{14}COOH$	Hexadecanoic	16:0	63
Stearic	$CH_3(CH_2)_{16}COOH$	Octadecanoic	18:0	70
Unsaturated fatty acids				
Oleic	$CH_3(CH_2)_7CH{=}CH(CH_2)_7COOH$		$18{:}1^{\Delta,9}$	13
Linoleic	$CH_3(CH_2)_4CH{=}CHCH_2CH{=}CH(CH_2)_7COOH$		$18{:}2^{\Delta,9,12}$	−5
Linolenic	$CH_3CH_2CH{=}CHCH_2CH{=}CHCH_2CH{=}CH(CH_2)_7COOH$		$18{:}3^{\Delta,9,12,15}$	−11
Arachidonic	$CH_3(CH_2)_4(CH{=}CHCH_2)_3CH{=}CH(CH_2)_3COOH$		$20{:}4^{\Delta,5,8,11,14}$	−50

* *Meaning of symbol* (e.g.)

$$18{:}2^{\Delta,9,12}$$

No. of C atoms ↓ — Position of double bonds (9–10 and 12–13) ↑ No. of double bonds

Groups attached to the carbon atoms joined by a double bond cannot freely rotate so two *geometrical isomers* are possible depending on the spatial arrangement of these groups. In the case of the *cis isomer* the groups are the same side of the double bond, while the *trans isomer* has the groups on opposite sides of the double bond.

Natural fatty acids are usually in the *cis* form and the *trans* isomer is comparatively rare.

$$CH(CH_2)_7CH_3 \qquad\qquad CH_3(CH_2)_7CH$$
$$\|\qquad\qquad\qquad\qquad\qquad\qquad \|$$
$$CH(CH_2)_7COOH \qquad\qquad CH(CH_2)_7COOH$$

cis-Δ^9-octadecenoic acid   *trans*-Δ^9-octadecenoic acid
(oleic acid)   (elaidic acid)

TRIACYLGLYCEROLS

The most abundant lipids in living organisms are the triacylglycerols or *triglycerides*. These are simple lipids which are fatty acid esters of the trihydric alcohol glycerol (Fig. 3.5). If the fatty acids substituted at positions 1 and 3 are different, then carbon atom number 2 becomes a chiral centre giving rise to D and L stereoisomers and the usual form in nature is the L isomer.

Triglycerides are called *neutral fats* as, unlike some other lipids, the molecules do not carry a charge. They are also known as *fats* or *oils* depending on whether they are solid or liquid at room temperature. The naturally occurring fats and oils are quite complex mixtures of triacylglycerols because of the wide variety of fatty acids that can be substituted at the three alcohol groups of glycerol.

Melting points   The melting point of fatty acids rises with increasing chain length and falls with the degree of unsaturation (Table 3.3). The physical state of natural fat therefore depends on the relative composition of the fatty acids that are present in the triacylglycerol. Beef suet, for example, is more solid than butter at room temperature due to the higher proportion of longer chain fatty acids. Fats from animals are generally solid at room temperature, while those from plants and fish are usually liquid because of their greater content of unsaturated fatty acids (Fig. 3.6).

Biological role

DIGESTION AND ABSORPTION

The triacylglycerols which account for the greater part of ingested lipid are hydrolysed in the gut to 2-monoacylglycerol and the sodium salts of the fatty acids (Fig. 3.5). The

$$CH_2OCOR_1 \qquad\qquad\qquad\qquad CH_2OH \qquad R_1COO^- + H^+$$
$$|\qquad\qquad\qquad\qquad\qquad\qquad\qquad\qquad\qquad |$$
$$R_2COO-C-H + 2\,H_2O \xrightarrow{\text{lipase}} R_2OCO-C-H +$$
$$|\qquad\qquad\qquad\qquad\qquad\qquad\qquad\qquad\qquad |$$
$$CH_2OCOR_3 \qquad\qquad\qquad\qquad CH_2OH \qquad R_3COO^- + H^+$$

Figure 3.5 The digestion of triacylglycerols by pancreatic lipase. Bile salts are required to form an emulsion with the triglycerides and their products of hydrolysis. Fatty acids are present in the emulsion as their sodium soaps.

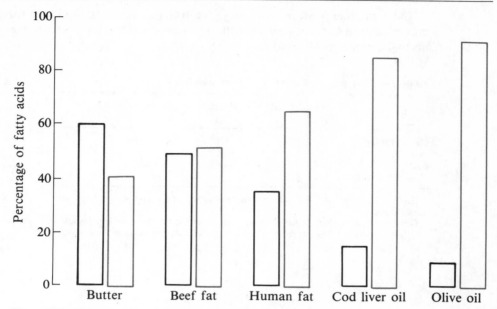

Figure 3.6 Fatty acid composition of some natural fats (saturated fatty acids, black; unsaturated fatty acids, red).

hydrolysis is catalysed by pancreatic lipase and aided by the emulsifying action of the bile salts. The products are then absorbed and the fats reformed in the gut mucosa which enables the composition of the triacylglycerols to be adjusted to that required by the animal. The intact esters then pass via the lymphatic system into the blood where they are transported as *chylomicrons*: fine droplets of triglyceride stabilized by a thin hydrophilic coat of lipoprotein.

OXIDATION AND STORAGE
Some of the triacylglycerols are broken down, and the glycerol and fatty acids are then oxidized in the tissues to provide energy. Any excess is stored as large droplets in specialized cells of the connective tissue known as *adipocytes*. These cells make up the adipose tissue which acts as a heat insulator and gives mechanical protection to vital organs as well as providing a long-term store of energy.

ENERGY VALUE OF FAT STORES
Fats are very reduced molecules and 1 mol of fat can store more than twice as much energy as 1 mol of carbohydrate. Triacylglycerols are also very hydrophobic so fat is stored in a virtually anhydrous form, unlike carbohydrate which is hydrophilic and therefore very hydrated. This means that fat is a very concentrated form of energy since 1 g of anhydrous triglyceride stores six times as much energy as 1 g of hydrated glycogen.

The total energy stored as triacylglycerols in humans is sufficient for weeks or months during starvation whereas the glycogen stores last for less than one day under normal conditions (Table 3.4).

Table 3.4 The energy reserves of the human body

	Energy stored (kJ/g)	Weight (g)	Total energy (kJ)
Carbohydrate (glycogen)	17	320*	5 440
Fat (triacylglycerols)	40	8000†	320 000

* This weight is for a well-fed human being; starvation or exercise will deplete this considerably.

† The weight shown here is for a young lean male: the fat stores in females are higher and the total fat stored can be at least twice this amount in the case of overweight people.

3.4 Phospholipids: the structural elements of membranes

Chemical structure

Complete hydrolysis of a compound lipid yields at least one other component as well as the usual alcohol and fatty acids. Phospholipids are an example of this since hydrolysis gives an alcohol, fatty acids and another component which is usually a base. Phospholipids, chemically, are mixed esters of fatty acids and orthophosphoric acid. Unlike the triglycerides, they do not act as a source of energy but are the main structural elements of membranes. There are two main groups of phospholipids: the *phosphoglycerides* and the *sphingomyelins*.

PHOSPHOGLYCERIDES (GLYCEROL PHOSPHATIDES)

These are very abundant in living organisms and have a similar structure to the triglycerides. As with neutral fats, the alcohol moiety is glycerol and carbon atoms 1 and 2 are esterified with fatty acids but unlike the triglycerides, carbon atom number 3 is esterified with orthophosphoric acid and not with a long-chain carboxylic acid. This molecule is *phosphatidic acid*, the parent compound of the phosphoglycerides.

One of the free phosphoric acid groups of the phosphatidic acid is also esterified with another alcohol to give the complete structure of a phospholipid. The four commonest of these alcohols in the order of their relative abundance are: *choline*, *ethanolamine*, *serine* and *inositol* (Fig. 3.7). The common names of the phosphoglycerides are derived from the nature of the alcohol linked to the phosphoric acid residue and the name of the alcohol is put first followed by the word phosphoglyceride. Alternatively, the prefix *phosphatidyl* is used before the name of the alcohol. For example, a phosphoglyceride containing serine is known as *serine phosphoglyceride* or *phosphatidyl serine*.

The second carbon atom of the glycerol forms a chiral centre and the natural

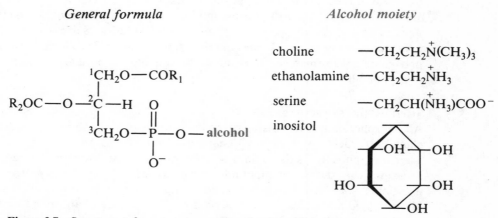

Figure 3.7 Structures of some common phosphoglycerides.

phosphoglycerides are of the L configuration, being related structurally to L-glyceraldehyde.

SPHINGOMYELINS

The sphingolipids are based on the alcohol *sphingosine* and not glycerol. In the case of the *sphingomyelins*, the primary alcohol group is esterified with phosphoric acid linked to either choline or ethanolamine. Sphingomyelins also contain a fatty acid linked to the amino group as an acyl derivative while the secondary alcohol remains unsubstituted. Sphingomyelins are present in membranes and are particularly rich in *myelin*, the fatty sheath which surrounds many nerves and acts as an electrical insulator.

$$CH{=}CH(CH_2)_{12}CH_3$$
$$|$$
$$CHOH$$
$$|$$
$$CHNH_2$$
$$|$$
$$CH_2OH$$

sphingosine

$$CH{=}CH(CH_2)_{12}CH_3$$
$$|$$
$$CHOH$$
$$|$$
$$CHNHOCR_1$$
$$\qquad\quad O$$
$$|\qquad\ \ \|$$
$$CH_2O{-}P{-}CH_2CH_2N^+(CH_3)_3$$
$$|$$
$$O^-$$

a sphingomyelin

Physical properties

ELECTRICAL CHARGE

Phospholipids are charged molecules in contrast to the neutral fats (Fig. 3.7). The phosphoric acid residue carries a negative charge while the nitrogen atom present in three out of the four common phospholipids is positively charged so these molecules are zwitterions. The fourth phospholipid has inositol as the alcohol moiety which is uncharged so that phosphatidyl inositol is not a zwitterion. Some phospholipids carry a further negative charge if they contain a serine residue. The alcohol attached to the phosphate group therefore determines the sign and magnitude of the charge on the phospholipid molecules.

AMPHIPATHIC MOLECULES

All phospholipids have a distinctly polar region which is hydrophilic, while the long-chain fatty acids and the hydrocarbon chain of the alcohol are hydrophobic. Therefore, phospholipids contain both hydrophilic and hydrophobic regions in the same molecule and such compounds are said to be *amphipathic*. For reasons discussed in Section 2.2, the hydrocarbon chains tend to associate with each other so that phospholipid molecules have a characteristic 'clothes peg' shape with the head as the polar region (Fig. 3.8).

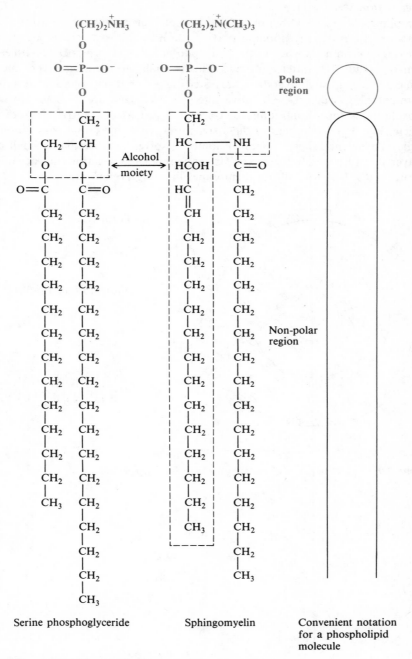

Figure 3.8 Phospholipids as amphipathic molecules.

INTERACTION WITH WATER

Phospholipids are insoluble molecules but interact with water to form *micelles*. In these structures, the hydrophilic part of the molecule seeks to associate with the aqueous environment, whereas the hydrophobic side chains associate with each other so as to exclude the water molecules. This is similar to the monolayers and micelles formed by a soap such as sodium stearate (Fig. 2.6) but phospholipids can also form more complex structures. The most interesting of these are *liposomes* in which the phospholipids form continuous bilayers arranged as concentric rings of a closed vesicle (Fig. 3.9). Treatment of these structures with ultrasound (sonication) gives a single bilayer enclosing a region of aqueous solvent, rather like a small cell. This bilayer structure is in fact very similar to the type of association that occurs between natural phospholipids to form membranes in living organisms.

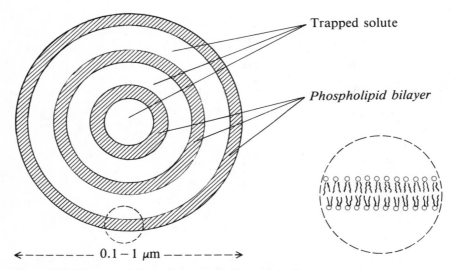

Trapped solute

Phospholipid bilayer

←——————— 0.1 − 1 μm ———————→

Figure 3.9   The association of phospholipids to form liposomes.

3.5 Steroids: bile salts and hormones

The structure of steroids

THE STEROID NUCLEUS

Steroids are derived lipids which are based on the 17 carbon allicyclic hydrocarbon compound perhydrocyclopentanophenanthrene.

perhydrocyclopentanophenanthrene

The steroid ring is virtually planar and substituent groups situated below this plane are marked as a dotted line (α). Similarly, those groups that are located above the plane are represented as a rigid line (β). The hydrogen atom at the C-5 position can be α or β and if it is α, the A and B rings are fused together in the *trans* position to give an almost planar structure. If the hydrogen is β, the A and B rings are in the *cis* position and form a buckled structure. The *trans* configuration of the A and B rings is present in the steroid hormones and the *cis* form in the bile salts. Methyl substituents on the rings are not usually written in as CH_3— but are shown simply as a bond (—).

Steroids are very hydrophobic compounds and are therefore insoluble in water but soluble in organic solvents. For this reason they are usually classified as lipids although many of them are not esters. The commonest group of steroids are in fact the *sterols* which are alcohols and therefore non-saponifiable.

CHOLESTEROL

Steroids are widespread in higher organisms where they perform a variety of functions and the most abundant of the steroids in animals is cholesterol. This occurs as the free alcohol and is also present as esters of long-chain fatty acids. Cholesterol is part of the structure of many cell membranes and is also the precursor of the bile salts and steroid hormones. Further details of its metabolism are considered in Part II.

cholesterol

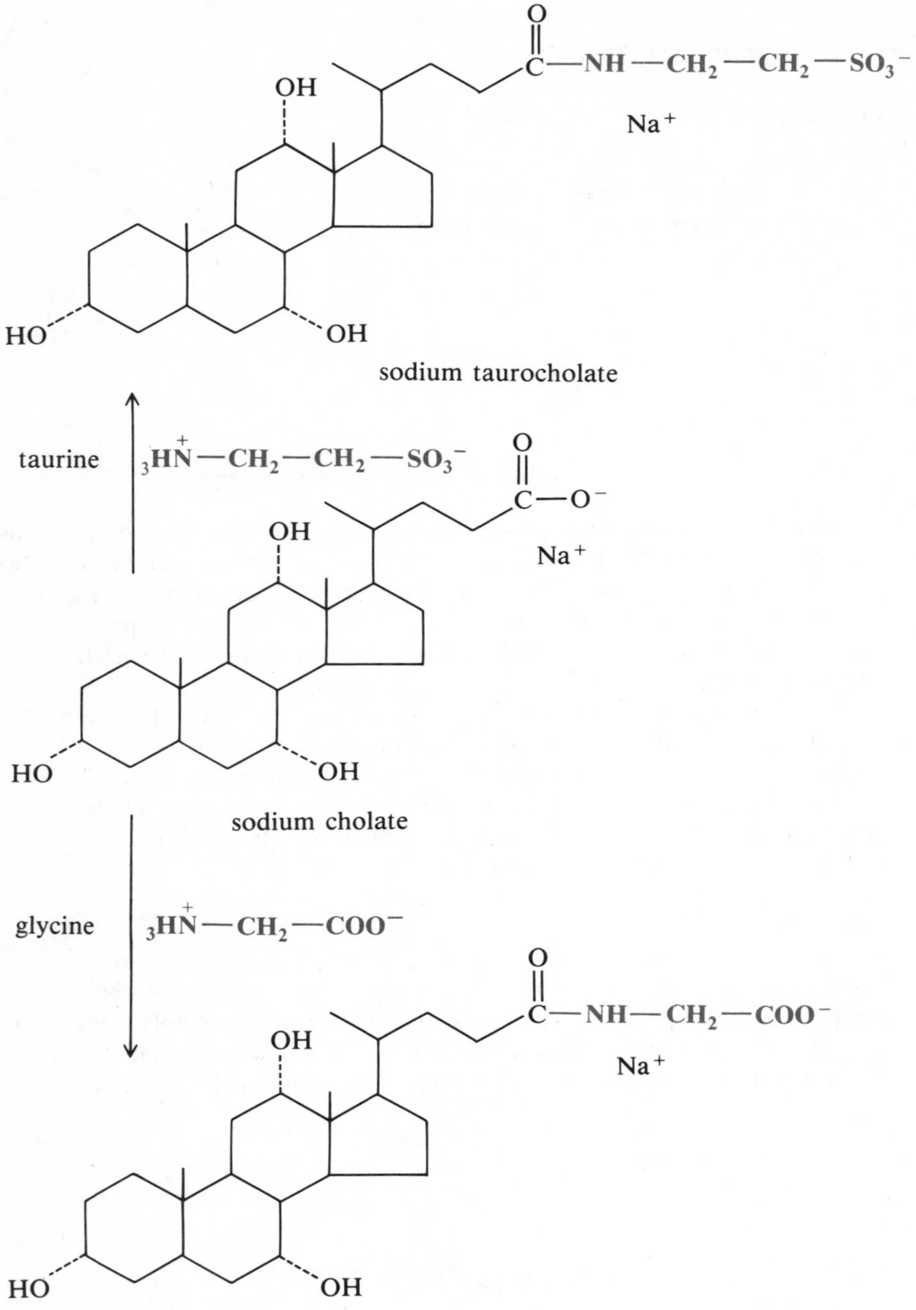

Figure 3.10   Bile salts. The conjugated amino acid residues (shown in red) and the hydroxyl groups are hydrophilic; the rest of the molecule is hydrophobic.

Bile salts and steroids

THE BILE ACIDS

One of the derivatives of cholesterol is *cholic acid*, the most abundant of the bile acids which form the basis of the bile salts in man. The carboxyl group of the cholic acid forms a peptide bond with the amino acids glycine or taurine to give the bile salts *sodium glycocholate* or *sodium taurocholate* (Fig. 3.10). The steroid ring is very hydrophobic while the taurine or glycine residues are hydrophilic; therefore, bile salts are amphipathic molecules and it is this property which forms the basis of their biological role.

Bile salts are synthesized in the liver and form a major component of the bile which is stored in the gall bladder. Bile is released into the small intestine following a fatty meal and the bile salts, because of their amphipathic properties, act as powerful emulsifying agents on the ingested fat. The bile salts thereby enable the fats to be more readily hydrolysed by enzymes and also greatly aid their absorption from the gut. In some diseases, where there is a deficiency of bile salts, undigested triglyceride appears in the stool together with the fat-soluble vitamins A, D, E and K.

STEROID HORMONES

Cholesterol is also the metabolic precursor of the steroid hormones which are active at well below micromolar (10^{-6} mol/litre) and even down to picomolar (10^{-12} mol/litre) concentrations. Hormones are secreted by the endocrine glands and act by binding to receptors on their target cells. These receptors are specific proteins or glycoproteins which have a high affinity and binding capacity for a particular hormone. Hormones that are polypeptides bind to receptors on the plasma membrane and affect transport processes and enzyme activities. Steroid hormones, on the other hand, bind to receptors inside the cell and enter the nucleus where they affect gene expression.

The effect of a particular hormone is very dependent on the chemical structure and quite small changes in structure can have a profound effect on physiological activity. There are many examples of this but one of the most striking is to be found with the sex hormones. The male sex hormone testosterone and the female sex hormone progesterone have very similar structures but widely differing biological activities (Table 3.5).

There are a very large number of steroid hormones, each one of which may show more than one physiological effect. The structures of some of the important steroid hormones selected from a number of groups are given in Table 3.5, together with a brief summary of their principal metabolic actions.

Table 3.5   Some examples of steroid hormones
(The example selected is shown in capital letters with the main class in parenthesis)

Steroid hormones	*Principal metabolic effects*

CORTISOL
(glucocorticoid)

Cortisol, synthesized by the adrenal cortex, has the opposite effect to insulin on metabolism by preventing the uptake of glucose by the cells. It increases blood glucose and stimulates the breakdown of fat.

Cortisol also reduces the body's immunity and is an effective anti-inflammatory agent.

ALDOSTERONE
(mineralocorticoid)

Aldosterone is synthesized by the adrenal cortex and regulates the water and electrolyte balance of the animal. It does this by stimulating the uptake of Na^+ and the excretion of K^+ by the kidney.

TESTOSTERONE
(androgen)

Testosterone, which is synthesized in the testes, is the principal male sex hormone. It is responsible for the development and maintenance of the male sex organs and the secondary sexual characteristics.

Testosterone together with the other androgens has an anabolic effect on metabolism and promotes the growth of muscle and bone.

| *Steroid hormones* | *Principal metabolic effects* |

PROGESTERONE
(progestin)

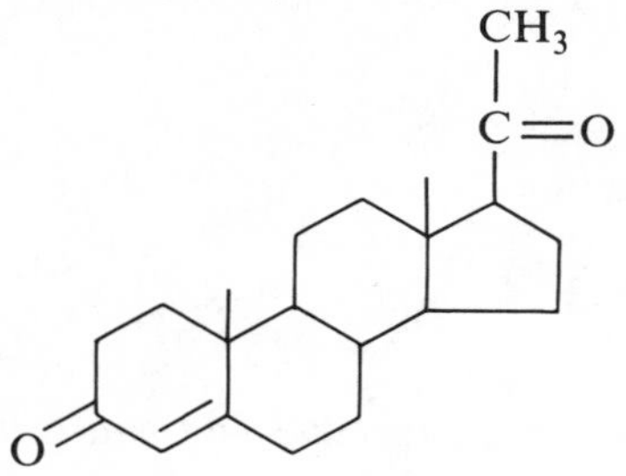

The female sex hormone progesterone is synthesized in the corpus luteum of the ovary.

It prepares the uterus to receive the fertilized egg and plays an important part in the maintenance of pregnancy.

OESTRADIOL
(oestrogen)

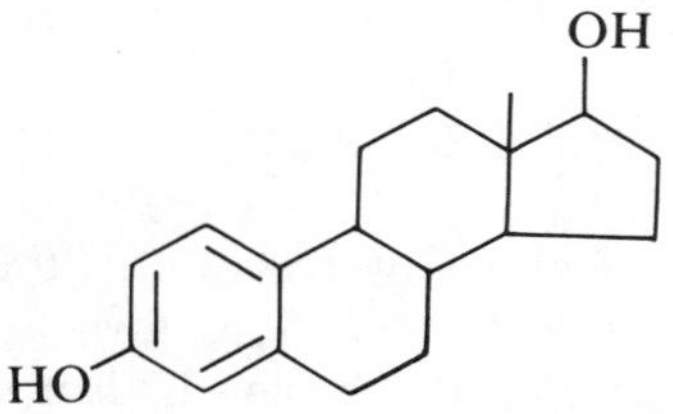

The other principal female sex hormone is oestradiol which is synthesized in the ovaries and placenta.

The hormone is responsible for the growth and maintenance of the female sex organs and secondary sexual characteristics. Together with progesterone it is responsible for the regulation of the menstrual cycle and the maintenance of pregnancy.

4. Macromolecules

4.1 Storage polysaccharides: the short-term food stores

Polymers of monosaccharides

THE GLYCOSIDIC BOND

The carbonyl group which is present in all sugars is very reactive and can form hemiacetals or acetals with other hydroxylic compounds (**XOH**).

carbonyl group   hemiacetal   acetal

The characteristic ring structure of sugars is brought about by reaction of the carbonyl group with the secondary alcohol group on C-4 or C-5 to give an intramolecular hemiacetal where χOH is the rest of the molecule. The new hydroxyl group on C-1 is quite reactive; under the right conditions it can combine with a hydroxyl group of another sugar, with the elimination of water, to form a *glycosidic bond* (—**O**—) and a new disaccharide molecule:

glucose   maltose
[α-D-glucopyranose]   [α-D-glucopyranosyl-α-1,4-D-glucopyranose]

This process can be repeated many times and the name *oligosaccharide* is given to molecules containing from two to ten monosaccharide units linked by glycosidic bonds to form a chain.

POLYSACCHARIDES

These are very large molecules of anything from 10 to a 1000 or more monosaccharide units joined together by glycosidic bonds. If all the monosaccharides are identical, the polymers are known as *homopolysaccharides* and if there are two or more basic units the polymers are called *heteropolysaccharides*. A particular monosaccharide unit can form the basis of a number of different macromolecules. This is possible because of the different size of the molecules, the variable degree of branching and cross-linking and the nature of the glycosidic bond. This is nicely illustrated in the case of glucose which is the basic component of a number of polysaccharides. The two most abundant of these glucose polymers are starch, which is the main food store of plants, and glycogen, which is the carbohydrate reserve of animals. In both cases, the glycosidic bond between the glucose molecules is of the α configuration and the geometry of the link is such that the polysaccharide chain forms tightly packed coils. Such structures are ideal for the formation of the granules in which starch is stored in the cell.

Starch

OCCURRENCE

Starch is synthesized from metabolic intermediates formed by photosynthesis and stored as a reserve of food in large quantities in the seeds (cereals) and tubers (potatoes) of plants. The hydrated granules of starch in these structures usually contain two polysaccharides, *amylose* and *amylopectin*, and the relative proportions of these molecules depend on the source of the starch.

AMYLOSE

This molecule consists of D-glucose units linked by α-1,4 bonds in the form of a straight chain. The number of glucose residues present varies widely so the molecular weight of amylose can be anything from 1000 to 500000. The characteristic blue

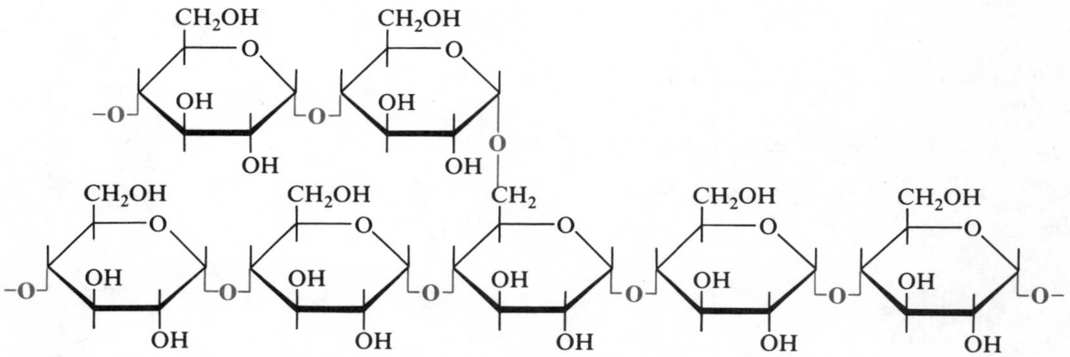

Figure 4.1 Part of the structure of amylopectin and glycogen. These molecules consist of a linear array of D-glucose units linked by α-1,4 glycosidic bonds and an α-1,6 branch point. Amylose has a similar structure but is unbranched.

colour formed when starch reacts with iodine is due to amylose which forms a hollow helix in which the iodine molecules become trapped.

AMYLOPECTIN

This molecule is similar to amylose but also contains α-1,6 links to give a branched structure. The molecule is very similar to glycogen but has fewer branch points (Figs. 4.1 and 4.2).

DIGESTION

The enzyme *α-amylase* in saliva and the gut catalyses the hydrolysis of α-1,4 bonds

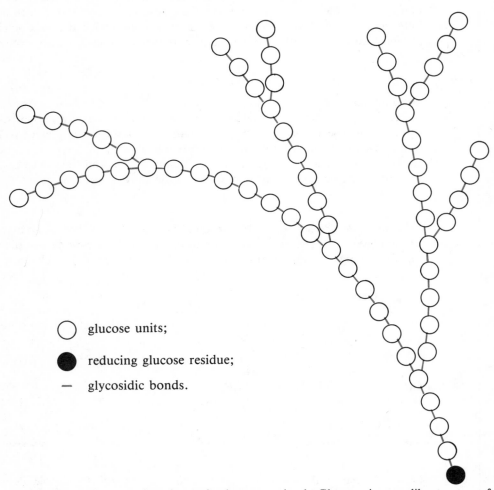

○ glucose units;

● reducing glucose residue;

— glycosidic bonds.

Figure 4.2 A representation of part of a glycogen molecule. Glycogen is a tree-like structure of α-1,4 linked linear arrays and α-1,6 branch points. Amylopectin is similar but is less branched.

throughout the starch molecule to give mainly maltose—a disaccharide of glucose—
and also some glucose and maltotriose—a trisaccharide of glucose. Amylose can be
hydrolysed completely by this enzyme but the α-1,6 bonds of amylopectin cannot be
broken down and the resulting *limit dextrin* resists further attack by the amylase once
the outer branches have been removed. The α-1,6 bonds on the limit dextrin are then
attacked by *α-1,6-glucosidase*—another enzyme—and the digestion is finally completed by pancreatic maltase which hydrolyses the di- and tri-saccharides to give
glucose.

Glycogen

OCCURRENCE

Glycogen is stored in a number of organs but the liver and the muscle are by far the
richest sources. The liver contains the greatest concentration of glycogen but the
skeletal muscle has the highest amount because of the greater mass of muscle
compared to liver (Table 4.1). The liver glycogen acts as the immediate source of
blood glucose while muscle glycogen provides a source of metabolic energy for use in
the muscle. Glycogen is present in these tissues in the form of granules (10–40 nm)
which also contain the enzymes required for the breakdown and synthesis of the
polysaccharide. The glycogen stores contain about 3400 kJ of energy which is enough
for 12 h or less depending on the activity undertaken.

Table 4.1   Glycogen stores in the liver and skeletal muscle of a well-fed man

	Weight of tissue (kg)	Glycogen concentration (% of wt)	Total glycogen (g)
Liver	1.6	5	80
Muscle	35	0.7	240

STRUCTURE

Glycogen has a similar structure to amylopectin with straight chains of α-1,4-linked
glucose units and α-1,6 branch points (Fig. 4.2) but the molecule is more branched and
compact than amylopectin. It is also a larger polysaccharide with molecular weights
as high as 10^6 to 10^8.

DIGESTION

The digestion of glycogen in the gut is the same as that of starch with glucose as the
final product.

4.2 Structural polysaccharides: extracellular molecules

Polysaccharides also play a key role in the structure of tissues in plants and animals and in the cell walls of micro-organisms. These polysaccharides are insoluble in water and this, together with their strength, is due to their chemical structure and make-up.

Homopolysaccharides

CELLULOSE

Carbohydrates, in general, are the most plentiful extracellular compounds in nature; cellulose, in particular, is the most abundant organic molecule on the planet. Its greatest occurrence is in the cell walls of plants which are made up of cellulose fibres set in a polymeric ground substance, rather like steel rods in reinforced concrete. The result is a very strong material that gives shape, support and protection to plants.

Structure   Cellulose, like starch and glycogen, is a homopolysaccharide of D-glucose but with the units linked β-1,4 instead of α-1,4 as in the storage polysaccharides; see Fig. 4.3. This small difference in the configuration of the glycosidic bond is responsible

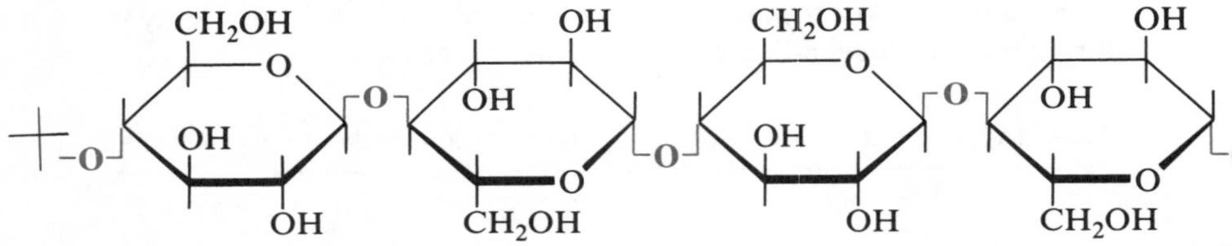

Figure 4.3   Part of the structure of a cellulose molecule showing the D-glucose units joined by β-1,4 glycosidic links (—O—).

for the characteristic physical and chemical properties of cellulose. The chains of the glucose units are unbranched and the geometry of the β links means that the polymer assumes an extended configuration rather than the typical coiled structure of the α-linked storage polysaccharides. Parallel molecules are held together by hydrogen bonds to form strong insoluble fibrils, and cross-linking of the molecules with aromatic residues, as in wood, further strengthens the cellulose to give a very tough material.

Digestion   The α-amylases present in the digestive tract of animals are unable to hydrolyse cellulose because of the stereochemistry of the β-glycosidic link. *Cellulases* capable of hydrolysing the β-glycosidic bond are found in the digestive tract of termites and slugs, which accounts for the destructive potential of these small animals on plant life. Cellulases are completely absent in mammals, and the herbivores such as cattle and sheep are only able to digest grass by reason of the cellulases present in the large number of bacteria found in the rumen.

CHITIN

The shells of crustaceans, such as crabs and lobsters, and the exoskeletons of insects consist largely of chitin—another homopolysaccharide. The basic unit is a sugar in which the hydroxyl group on the 2-position of D-glucose has been replaced with an acetyl amino group to give N-*acetyl-D-glucosamine* (Fig. 4.4). The substituted glucose

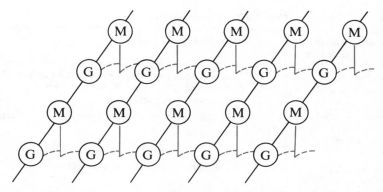

N-acetyl-D-glucosamine

(G)

O-lactyl-*N*-acetyl-D-glucosamine
(*N*-acetylmuramic acid)

(M)

small polypeptide

$_3$HN—Gly—Gly— Gly—Gly—Gly—COO$^-$

cross-linking peptide chain

(*a*) The components of murein (peptides as in *Staphylococcus aureus*).

(*b*) Part of the peptidoglycan molecule. The sugar components -(G)—(M)- are linked β-1, 4; the tetrapeptide is joined to the —COO$^-$ of muramic acid by a peptide bond and the whole structure cross-linked by a peptide chain from D-Ala to L-Lys.

Figure 4.4 The structure of murein, the peptidoglycan of bacterial cell walls.

units are linked β-1,4 to give a tough linear chain and, in the crustaceans, the matrix of chitin is further hardened by deposition of calcium carbonate.

Heteropolysaccharides and complex carbohydrates

ACID MUCOPOLYSACCHARIDES

These molecules are heteropolymers made up of a recurring disaccharide unit which consists of an aminohexose such as D-*glucosamine* joined to an acid sugar derivative, usually D-*glucuronic acid*.

β-D-glucosamine β-D-glucuronic acid

They all carry a negative charge because of the —COO⁻ groups and in some cases this charge is further enhanced by esterification of some of the secondary alcohol groups with $-SO_4^{2-}$. These acidic molecules were first isolated from *mucin*, the slippery lubricant of mucus, and were therefore called *acid mucopolysaccharides*. Today, they are more commonly known as the *glycosaminoglycans*. These large molecules are *polyelectrolytes* and associate with a large volume of water to form viscous and gelatinous compounds which act as lubricants at joints. Many of the

Table 4.2 Some common heteropolysaccharides

Glycosaminoglycan (monosaccharide components)	Physiological function
Hyaluronic acid (D-glucuronic acid + N-acetyl-D-glucosamine)	The major component of the intracellular cement of connective tissue and also of synovial fluid, the lubricant and shock absorber that surrounds joints.
Chondroitin sulphates (D-glucuronic acid + N-acetyl-D-galactosamine 4- or 6-sulphate)	Very abundant in cartilage and skin as part of proteoglycans.
Keratan (D-galactose + N-acetyl-D-glucosamine-6-sulphate or galactosamine-6-sulphate)	An important component of skin and cartilage.
Heparin (L-iduronic acid-2-sulphate or D-glucuronic acid + D-glucosamine-2,6-disulphate)	An anticoagulant present in the lining of arterial cell walls. Secreted into the blood by mast cells.

glycosaminoglycans are associated with protein to form large complex *proteoglycan* molecules such as those present in the intracellular cement. Some examples of the glycosaminoglycans and a brief note on their function in the body are given in Table 4.2.

GLYCOPROTEINS AND GLYCOLIPIDS

These are large complex molecules in which the protein or lipid component is the dominant part of the molecule. However, the minor oligosaccharide part confers important properties on the compound. For example, the membrane of animal cells contains from 2 to 10 per cent carbohydrate and the structure of this small component determines the antigenic properties, recognition, adhesion and contact inhibition of the cell: properties which play an important part in the development and growth of normal and cancer cells.

The blood group specificity (A, B, O) also depends on the nature of the terminal monosaccharide residue attached to a pentasaccharide chain of a membrane gly-colipid. The group A antigen has *N*-acetyl-D-glucosamine, group B has galactose and group O has no monosaccharide in the sixth position.

PEPTIDOGLYCANS

A third type of association between carbohydrate and polypeptides is found in *murein*, a peptidoglycan which is the major structural component of bacterial cell walls. The heteropolysaccharide component consists of repeating units of the disaccharide N-*acetyl-D-glucosamine-β-1,4*-N-*acetylmuramic acid*, arranged in the form of parallel chains. A small polypeptide containing D amino acids is attached to the *N*-acetylmuramic acid and the parallel chains are cross-linked by short peptide chains joining the small peptides. The whole structure thus forms a gigantic molecule that completely envelops the bacterium rather like an elastic net surrounding a continental sausage (Fig. 4.4). This and other molecules in the cell wall give a structural rigidity and shape to the bacterium and prevent the cell membrane from rupturing in an osmotically hostile environment. The actual composition of the peptides depends on the particular species of bacteria and the structure of the peptidoglycan shown in Fig. 4.4 is that found in *Staphylococcus aureus*.

4.3 Proteins: the basis of living matter

Occurrence

Polysaccharides are the commonest extracellular molecules in the biosphere but proteins are the most abundant macromolecules found within living cells where they can account for up to 50 per cent of the dry weight. Chemically they are polymers of hundreds or thousands of amino acids joined together by peptide bonds (Table 4.3).

Table 4.3 Proteins as macromolecules

Protein	Number of amino acid residues	Number of polypeptide chains	Molecular weight
Glucagon	29	1	3 500
Insulin	51	2	5 733
Myoglobin	153	1	16 890
Haemoglobin	574	4	64 500
Immunoglobulin	1320	4	160 000
Collagen	∼ 3000	3	300 000

Proteins are ubiquitous molecules that are essential to all forms of life and they well deserve the name which comes from the Greek *proteios* meaning of primary importance.

Proteins are also found in association with other molecules as *conjugated proteins*. For example, the *haem proteins* of oxygen transport contain porphyrin rings, the *flavoproteins* involved in oxidations are joined to flavin nucleotides and the *lipoproteins*, which transport cholesterol and triglycerides in the blood plasma, are associated with lipid.

Nutrition

THE DYNAMIC STATE OF PROTEINS

During digestion, proteins in the food are broken down to amino acids which are then absorbed and used to synthesize the proteins needed by the body. Labelling proteins with the heavy isotope ^{15}N has shown that these molecules are being continuously broken down and synthesized (Fig. 4.5) and in man the turnover is about 400 g per day. There is a large daily excretion of nitrogen from the body, which in mammals consists mainly of urea $CO(NH_2)_2$, and this nitrogen loss must be replaced. Protein is the main source of nitrogen in the diet and the average adult male requires about 60 g per day.

NITROGEN BALANCE

Healthy adults are in a state of nitrogen balance in which the nitrogen input in food is exactly balanced by the output in urine and faeces. In the case of *negative nitrogen*

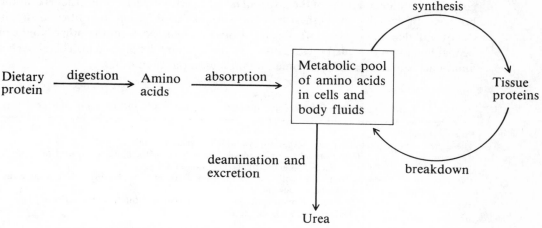

Figure 4.5 The dynamic state of proteins.

balance, the excretion exceeds the intake and this occurs with a poor diet and in diseases involving an excessive breakdown or loss of protein. Such a situation is very serious if it persists since protein, unlike carbohydrate or fat, cannot be stored to any great extent. A positive nitrogen balance in which the intake exceeds the excretion is normal in pregnancy, growing organisms and mature animals recovering from illness.

ESSENTIAL AND NON-ESSENTIAL AMINO ACIDS

The nutritional value of protein in the diet depends not only on how well it is digested and absorbed but also on its amino acid composition. The dietary proteins must

Table 4.4 The dietary requirements for amino acids in man

Essential	*Non-essential*
Arginine*	Alanine
Histidine	Asparagine
Isoleucine	Aspartic acid
Leucine	Glutamic acid
Lysine	Glutamine
Methionine	Glycine
Phenylalanine	Hydroxyproline
Threonine	Proline
Tryptophan	Serine
Valine	
	Cysteine†
	Cystine†
	Tyrosine‡

* Only essential for children.
† Only non-essential if sufficient methionine.
‡ Only non-essential if sufficient phenylalanine.

contain adequate amounts of the *essential amino acids*. These are amino acids that the body is either unable to synthesize or cannot make in sufficient quantities to supply the overall demand (Table 4.4). The *non-essential amino acids* on the other hand can be manufactured provided there is a sufficient overall intake of protein. Therefore, 'all amino acids are essential but some are more essential than others'.

Table 4.5 Some examples of the diverse functions of proteins found in living matter

Structural proteins

These are tough extracellular proteins that are insoluble in aqueous solution.

Collagen	This protein has a high tensile strength and is the main constituent of connective tissue, cartilage and animal hides.
Elastin	In contrast to collagen, elastin is an elastic protein and forms part of the walls of blood vessels and ligaments.
α-Keratins	These are the main components of external structures in animals such as feathers, nails, claws and horn.

Storage proteins

Proteins are not stored to any extent in animals and the following examples are exceptions to this general rule.

Ovalbumin	This protein, found in egg white, provides a store of food for the growth and development of the chick embryo.
Casein	This milk protein is the basic nutrient of infants and young animals.

Hormones

There are many examples of protein or peptide hormones and probably the best known of these are briefly mentioned below.

Glucagon	This hormone is produced by the cells of the pancreas and acts by increasing the blood glucose and stimulating the breakdown of fat.
Insulin	This well-known hormone is produced by the cells of the pancreas and has the opposite metabolic effect to that shown by glucagon.
Growth hormone	This hormone is responsible for the control of growth and development.
Erythropoietin	The production of erythrocytes is stimulated by erythropoietin and the hormone is very active following the loss of blood.

Transport proteins

A number of proteins, particularly those found in the blood plasma, transport materials throughout the body.

Haemoglobin	This is probably the best known of these proteins for its carriage of oxygen in the blood to all the tissues of the body.
Albumin	Albumin is the major component of the plasma proteins and acts as a carrier for a variety of substances including fatty acids, hormones and drugs.

Most of the world's population do not have an adequate intake of protein and the effects of this can be seen in many children of Third World countries who suffer from *kwashiorkor*. This is due to a chronic protein deficiency arising from a poor diet after weaning and the name comes from an African word meaning 'weaning disease'. The typical signs are poor growth, muscle wasting, oedema, malabsorption and susceptibility to infection. Even if the child survives by being given an adequate diet, it is likely that brain damage has occurred leading to permanent mental impairment. The importance of an adequate supply of protein is thus tragically demonstrated.

Function

The main function of proteins is to act as essential components of structural material, in contrast to that of carbohydrates and fats which is to provide energy. There are 22 different amino acids commonly found in proteins and the order of the amino acid residues determines the structure as discussed in the next section. The number of ways that the amino acids can be arranged in order is quite considerable when hundreds or thousands are involved. This is why the number of proteins found in nature is very large indeed and the human body alone is thought to contain somewhere in the region of 10 000 different proteins. These large numbers of different molecules perform an enormous range of structural and metabolic functions in living organisms, some of which are briefly considered below.

Some proteins are defensive molecules, for example the *immunoglobulins* which are *antibodies* manufactured by animals in response to foreign substances known as *antigens*. *Bacterial toxins* and *snake venoms* are also good examples of proteins produced to protect these organisms.

Muscle is largely protein and the interaction between *myosin*, the protein found in the thick filaments of skeletal muscle, and *actin*, the corresponding protein of the thin filaments, forms the basis of muscular contraction.

All enzymes are proteins and life would be impossible without these ubiquitous molecules which catalyse a vast number of reactions in living matter.

Finally, some more examples of the diversity of proteins are given in Table 4.5.

4.4 Proteins: their amino acid composition and structure

Primary structure

Proteins are polymers of amino acids linked together by peptide bonds and these large molecules show several levels of structural organization. The first of these is the *primary structure* which is the sequence of amino acids.

AMINO ACID COMPOSITION

The constituent amino acids of a protein can be identified and measured by hydrolysing the peptide bonds and separating the resulting small peptides and amino acids by chromatography and electrophoresis. Complete hydrolysis of the protein is achieved by strong acid or alkali and partial hydrolysis by the use of enzymes.

AMINO ACID SEQUENCE

The determination of the order in which the amino acid residues are arranged in the molecule is based on chemical methods that identify the free —COOH and —NH$_2$ groups. The protein is first hydrolysed by different enzymes that are specific for peptide bonds linking particular amino acids. This gives a series of smaller peptides whose structures can then be determined by identifying the C and N terminal amino acid residues and by further hydrolysis. The structures of the large peptides are then built up and the sequence of amino acid residues established by computer analysis. An example of the methods employed to determine primary structure is the use of phenylisothiocyanate or Edman's reagent (Fig. 4.6).

PEPTIDE BOND

The peptide bond linking the amino acids is planar due to some double bond character arising from the presence of resonance forms and the side chains of the amino acids are usually *trans* to the polypeptide backbone (Fig. 4.6).

Secondary structure

The planar peptide bonds can twist about the α-carbon atom so that the polypeptide chain can take up different arrangements in space and this spatial configuration of the polypeptide chain is the *secondary structure*. A large number of structures are theoretically possible but only a limited number are found in nature.

α-HELIX

In this structure, the polypeptide chain forms a helix with 3.6 amino acid residues per turn and a pitch of 0.54 nm. The helical shape is maintained by intramolecular hydrogen bonds (———) between the carbonyl oxygen and the amide hydrogen of peptide bonds that are three residues apart. Each hydrogen bond is quite weak but the large number involved gives a very stable structure.

$$\backslash C=O --- H—N /$$

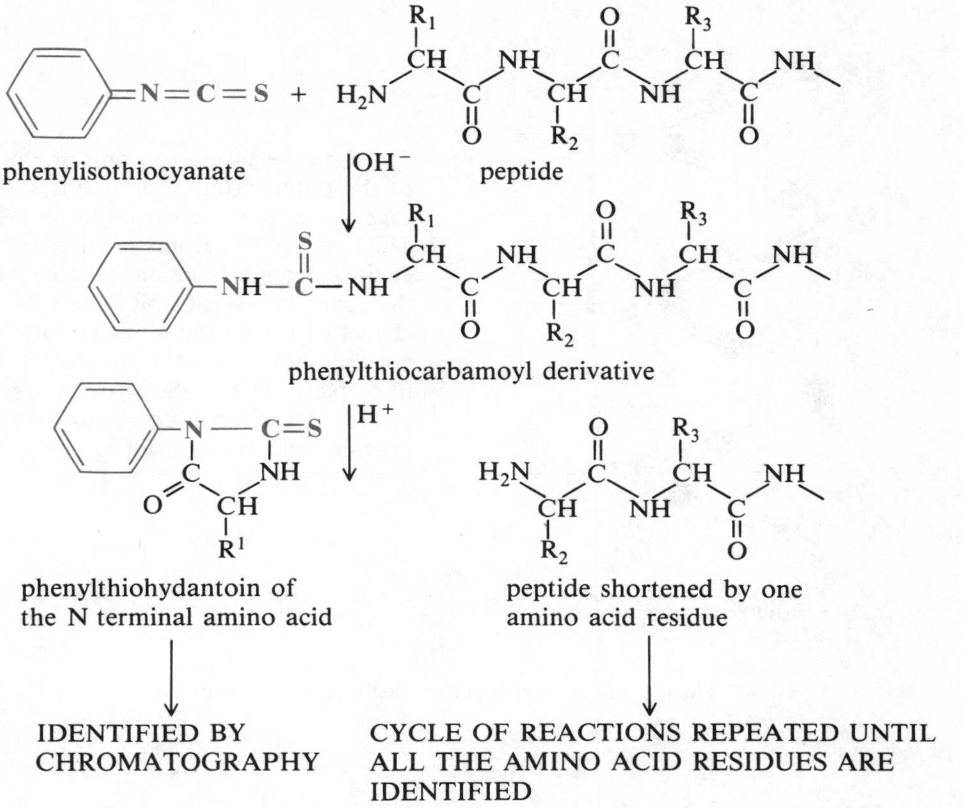

Figure 4.6 The determination of the sequence of amino acids in a peptide chain using Edman's reagent.

The amino acid side chains can be readily accommodated in the α-helix as they stick out into space away from the polypeptide chain (Fig. 4.7).

β-PLEATED SHEET
Another form of secondary structure is the β-pleated sheet in which the hydrogen bonds stabilizing the structure are intermolecular rather than intramolecular. The bonds occur between two parallel polypeptide chains which give an overall pleated shape to the structure. The adjacent polypeptides may be parallel, where the chains run in the same direction with the nitrogen atoms at the same end, or they may be antiparallel, where adjacent chains run in opposite directions (Fig. 4.8).

RANDOM COIL
The third type of configuration is the random coil where there is no consistent relationship between the amino acids and their stabilizing bonds.

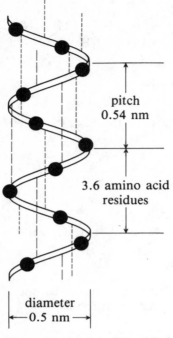

pitch
0.54 nm

3.6 amino acid
residues

diameter
←—0.5 nm —→

For the sake of clarity, only the line of the polypeptide backbone is shown, together with the α-carbon atoms ●. The radii of the atoms in the peptide chain are quite large and almost fill the space inside the coil. The side chains of the amino acid residues which are not shown point away from the helix into space. The whole structure is stabilized by hydrogen bonds between peptide links three residues apart,
– – – – –

Figure 4.7   Model of the right-handed α-helix present in proteins.

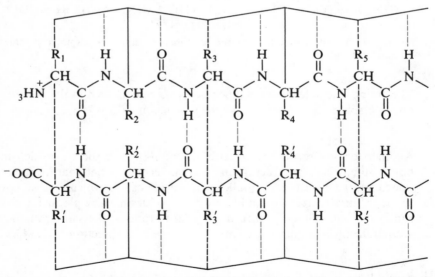

Figure 4.8   The structure of an antiparallel β-pleated sheet. The hydrogen bonds linking the chains are shown as broken red lines.

Tertiary structure

The next order of structure is the *tertiary structure* which is the folding of the polypeptide chain and the three-dimensional arrangement of the protein in space. In the case of *globular proteins*, the overall shape is compact while *fibrous proteins* form long thin threads.

INTRAMOLECULAR INTERACTIONS

The unique folding of the molecule in space is determined solely by the sequence of amino acids and is stabilized by various intramolecular interactions (Fig. 4.9).

Covalent bonds   The disulphide link is formed from two adjacent cysteine residues and is a strong covalent bond that confers considerable rigidity on the protein molecule.

Ionic bonds   The next strongest interaction is the electrostatic or ionic bond formed when an acidic amino acid residue (Asp, Glu) lies close to a basic amino acid residue (Lys, Arg) in space.

Hydrogen bonds   Numerous hydrogen bonds can be formed between the side chains of a number of amino acid residues (Asp, Glu, His, Ser, Thr). They are individually weak but because there are so many they make a big contribution to the structure.

Hydrophobic bonds   As discussed earlier, these bonds arise from a tendency of certain groups to exclude water molecules from their vicinity. As with the hydrogen bond,

Peptide chain

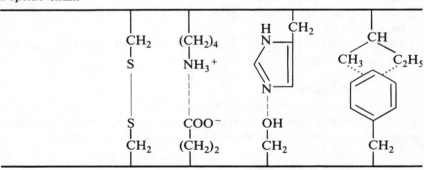

Peptide chain

Type of bond	covalent	ionic	hydrogen	hydrophilic
Bond strength	strong	moderate	weak	very weak

Figure 4.9   Some examples of the molecular interactions responsible for the maintenance of the tertiary structure of proteins.

each hydrophobic interaction is weak but the large number confers considerable stability on the structure.

AMINO ACIDS AND TERTIARY STRUCTURE

Generally speaking, the *hydrophobic amino acids* are found buried deep within the molecule away from the aqueous environment, whereas the polar *hydrophilic amino acids* tend to associate with water and are present on the surface of the protein. Fundamentally then, water is the driving force for the folding of the molecule.

Other amino acids may have more specific effects on the structure. For example, the two —SH groups of adjacent *cysteine* residues may form a —S—S— bond which confers considerable rigidity to the molecule. In addition, if the nitrogen of the imino group, present in *proline* and *hydroxyproline*, is involved in the peptide bond then the adjacent amino acid side chains are *cis* instead of the usual *trans* configuration. Such a structure cannot be accommodated into an α-helix and, where proline occurs, there is a marked change in the direction of the polypeptide chain.

Quaternary structure

Some proteins consist of more than one polypeptide chain and the association of these *subunits* in the protein gives rise to the quaternary structure. Probably the best known example of a protein with quaternary structure is haemoglobin which in the adult human consists of two α and two β subunits.

Many enzymes that have quaternary structure are *allosteric* and play an important role in metabolic control and this is discussed later.

HIGHER ORDERS OF STRUCTURE

In some cases, several proteins may associate to give complex structures and some examples of this are given below.

MICROTUBULES

Two proteins, α- and β-tubulin, form microtubules which are the supportive elements of the cell.

MULTIENZYME COMPLEXES

Sometimes a group of enzymes are associated together to form a multienzyme complex which catalyses a sequence of metabolic reactions. An example of this is the fatty acid biosynthesis complex which consists of seven enzymes.

MACROMOLECULAR ASSEMBLIES

Even higher orders of structure are possible and the *ribosomes* in *Escherichia coli* are a complex of 55 different proteins and three RNA molecules.

4.5   Nucleic acids: the blueprints for survival

Function

Nucleic acids are another important group of nitrogen-containing compounds and there are two main classes of these macromolecules: *deoxyribonucleic acid* (*DNA*) and *ribonucleic acid* (*RNA*).

DEOXYRIBONUCLEIC ACID

This is a very large molecule with a highly concentrated store of information which determines the properties of the cell and the way it grows and divides. DNA is the genetic material of the cell and in bacteria the molecule is present in the form of a single strand, whereas in higher organisms the DNA is associated with basic proteins called *histones* and some RNA to form the *chromosomes*. The number of chromosomes and therefore the amount of DNA is the same in all the cells of a particular species, with the exception of the germ cells which contain half the amount of DNA. In eukaryotic cells, virtually all of the DNA is located in the nucleus but trace amounts are also found in chloroplasts and mitochondria.

RIBONUCLEIC ACID

There are four main types of RNA and these are distributed throughout the cell. Approximately 70 per cent is present as *ribosomal RNA* (*rRNA*) and this is found in ribosomes which are small particles of ribonucleoprotein present in the cytoplasm of the cell. A further 15 per cent is present as soluble or *transfer RNA* (*tRNA*) and about 5 per cent is present as *messenger RNA* (*mRNA*). Some RNA is found in the nucleus of the cell (10 per cent) and there are also trace amounts in the mitochondria. The different types of RNA, together with the nuclear DNA, form an elaborate system for synthesizing proteins.

THE PROGRAM OF THE CELL

The relationship between the nucleic acids and the cell can be likened to a program which stores information and gives instructions to a computer. DNA is the store of genetic information and is also the template for RNA which determines protein synthesis. These processes are complex and can only be understood by knowing the composition and structure of the nucleic acids. The chemistry of these molecules is considered in this section and the details of the replication of DNA and the mechanism of protein synthesis are discussed in Chapter 11, 'Molecular biology'.

Chemical composition

NUCLEOTIDES

Controlled hydrolysis of DNA and RNA gives nucleotides which can be regarded as the basic units of nucleic acids just as monosaccharides are the building blocks of polysaccharides and amino acids of proteins. Nucleotides contain three components: a base, a sugar and a phosphate group. The 1'-position of the sugar is linked to a nitrogen atom (C-1 for pyrimidines, C-9 for purines) and one of the free —OH groups of the pentose is esterified with phosphoric acid. All the free hydroxyl groups can be esterified but the 5′-position is the commonest.

NUCLEOSIDES

Further careful hydrolysis of the nucleotides removes the phosphate group and leaves a nucleoside which is a base linked to a sugar. Complete hydrolysis of the nucleoside then gives a sugar and a range of bases.

The pentose sugars

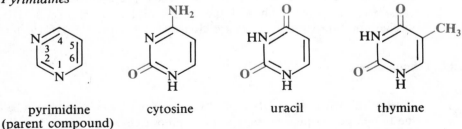

β-D-ribofuranose   β-D-2-deoxyribofuranose

Purines

purine
(parent compound)   adenine   guanine

Pyrimidines

pyrimidine   cytosine   uracil   thymine
(parent compound)

Figure 4.10 The structure of the sugars and main purines and pyrimidines present in nucleic acids.

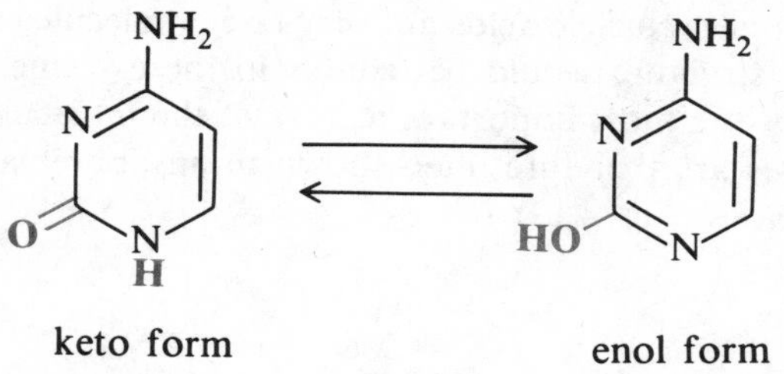

keto form   enol form

Figure 4.11 The keto and enol forms of cytosine.

BASES

Chemically, there are two groups of bases, the purines and the pyrimidines (Fig. 4.10), and both types exhibit tautomerism so the molecules can exist in the keto or enol forms (Fig. 4.11). Adenine, guanine and cytosine occur in both types of nucleic acids but thymine is only present in DNA while uracil is only found in RNA.

SUGARS

The other difference in composition of the nucleic acids is in the pentose sugars which give the names to the two types of nucleic acids: deoxyribose is in DNA while ribose is the sugar of RNA (Fig. 4.10).

NOMENCLATURE

The nucleosides are named after the base that is present. If the pentose is deoxyribose then the prefix deoxy is used (Table 4.6).

Table 4.6 The nomenclature of nucleosides and nucleotides

Base	Nucleoside	Nucleotide
RNA		
Adenine (A)	Adenosine	Adenosine-5′-monophosphate (AMP)
Guanine (G)	Guanosine	Guanosine-5′-monophosphate (GMP)
Cytosine (C)	Cytidine	Cytidine-5′-monophosphate (CMP)
Uracil (U)	Uridine	Uridine-5′-monophosphate (UMP)
DNA		
Adenine (A)	Deoxyadenosine	Deoxyadenosine-5′-monophosphate (dAMP)
Guanine (G)	Deoxyguanosine	Deoxyguanosine-5′-monophosphate (dGMP)
Cytosine (C)	Deoxycytidine	Deoxycytidine-5′-monophosphate (dCMP)
Thymine (T)	Deoxythymidine	Deoxythymidine-5′-monophosphate (dTMP)

Macromolecular structure

PRIMARY STRUCTURE

Nucleic acids are formed when the phosphate group of one nucleotide is joined to the sugar of the next nucleotide by an ester bond in the 3′-position, thus forming a long chain known as a *polynucleotide*. Nucleic acids are very large molecules and to draw anything like the complete structure would be tedious in the extreme. The actual sequence of bases is perhaps the most important feature of the molecule, as will be seen later, and the structures are therefore often shown in an abbreviated form or simply as the sequence of bases (Fig. 4.12).

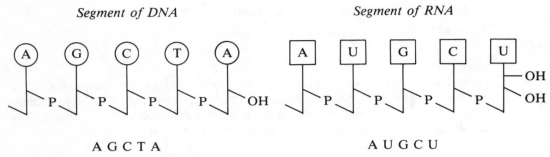

Segment of DNA

Segment of RNA

A G C T A   A U G C U

Figure 4.12   Shorthand notation for oligonucleotides.

SECONDARY STRUCTURE OF DNA

This nucleic acid is a long thin fibrous molecule with a molecular weight as high as 10^9. It consists of two polynucleotide chains held together by hydrogen bonds between pairs of bases (Fig. 4.13). The two chains are *antiparallel* and are in the form of a spiral or *double helix* with the 5'- and 3'-sequences running in opposite directions. This model was proposed by Watson and Crick in 1953 and is now the accepted structure for DNA. Most DNA molecules have this type of structure and this forms the basis for explaining the replication of DNA.

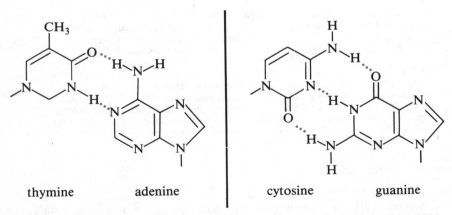

thymine   adenine   cytosine   guanine

Figure 4.13   The pairing of bases by hydrogen bonding (broken red lines) in DNA.

SECONDARY STRUCTURE OF RNA

Unlike DNA, the RNA molecule is generally a single random coil although there are some helical regions which show base pairing. Messenger and ribosomal RNAs have molecular weights of a 10^6 or more but the tRNAs are much smaller with molecular weights in the region of 25 000. In the case of the tRNAs, the secondary structures are known and these play an important part in protein synthesis, as discussed later in the book.

5. The organization of living cells

5.1 Membranes: the molecular boundaries

Membranes are thin, flexible boundaries that surround living cells and also divide them into distinct compartments or regions. They were first postulated as a functional necessity in the nineteenth century when it was observed by light microscopy that cells behave as osmometers by swelling in hypotonic media and shrinking in hypertonic solutions as if surrounded by a semi-permeable membrane. However, it was not until the twentieth century that membranes were visualized by electron microscopy and found to be only 6–10 nm thick. Today, some membranes can be isolated in a pure state and their chemical composition determined. Furthermore, specialized physical studies of membranes using techniques such as spectroscopy, nuclear magnetic resonance and electron spin resonance have enabled biochemists to see how the various components are related in the overall structure and this has helped to explain many of the properties of membranes.

Chemical composition

Membranes are large structures that contain lipid and protein as their main components, together with a small amount of carbohydrate material. The ratios of lipid to protein depend on the source of the membrane and range from 4:1 in myelin in nerve cells to 1:3 in bacterial cell membranes although many membranes have a similar lipid to protein ratio to those isolated from human erythrocytes. The overall chemical composition of some isolated membranes is shown in Fig. 5.1.

LIPID

This material forms the basis or foundation of membranes and consists mainly of *phospholipids* which are always present. Many membranes also contain *sterols* and in animal cells these are principally in the form of cholesterol which may be free or esterified. A third group of minor but important components are the *glycolipids*. The relative amounts of these lipid classes, as well as the proportion of individual lipids, is constant for a particular membrane but the composition of the fatty acid component can vary widely depending on the diet and other factors.

PROTEIN

Membrane proteins are generally hydrophobic and usually need organic solvents to bring them into solution. Solubilization therefore often leads to a change in the conformation of the proteins and a loss in their biological activity. The isolated

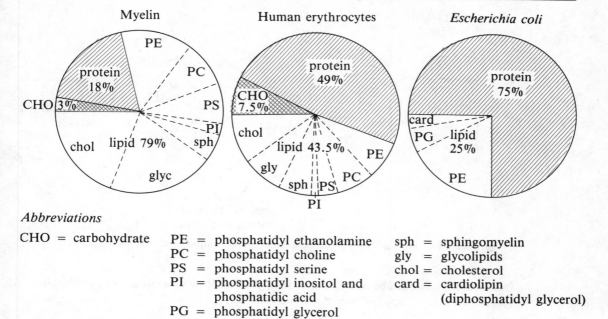

Abbreviations

CHO = carbohydrate

PE = phosphatidyl ethanolamine
PC = phosphatidyl choline
PS = phosphatidyl serine
PI = phosphatidyl inositol and
   phosphatidic acid
PG = phosphatidyl glycerol

sph = sphingomyelin
gly = glycolipids
chol = cholesterol
card = cardiolipin
   (diphosphatidyl glycerol)

Figure 5.1 The chemical composition of some membranes.

proteins tend to aggregate readily in aqueous solutions which makes them difficult to work with. For this reason they are less well understood than the membrane lipids, although some proteins have been isolated and studied in detail.

Membrane proteins contain a high proportion of hydrophobic amino acids and a relatively high percentage of acidic amino acids (Asp, Glu). They also generally have a low content of cysteine and so contain few if any —S—S— bonds which makes them quite flexible. Membrane proteins also have a high α-helical content.

CARBOHYDRATE

This is a minor component which is found in many but not all cell membranes and is present on the cell surface where it is responsible for the characteristic antigenic properties of the cell and also for most of its negative electrical charge.

Membrane structure

LIPID BILAYER

If lipid is extracted from cell membranes and spread as a monolayer at an air–water interface, then the area occupied by the lipid is found to be twice that of the area of the cell. This led to the suggestion that the lipid exists as a bilayer round cells, with the hydrophobic parts of the component molecules associated with each other and the hydrophilic parts exposed to the aqueous environment (Fig. 5.2). This bilayer

Peripheral (extrinsic)
membrane protein

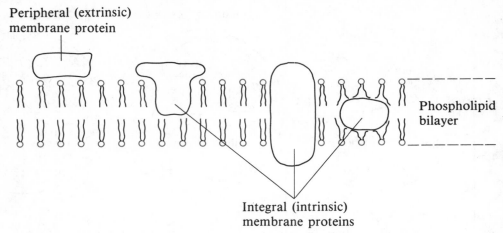

Phospholipid
bilayer

Integral (intrinsic)
membrane proteins

Figure 5.2   The association of proteins with the phospholipid bilayer of membranes.

structure can be seen by electron microscopy which shows membranes as a trilaminar structure when stained with OsO_4.

THE FLUID MOSAIC MODEL
The idea of the lipid bilayer was accepted early on but controversy continued about the exact location of the protein in the membranes. The model for membrane

Proteins

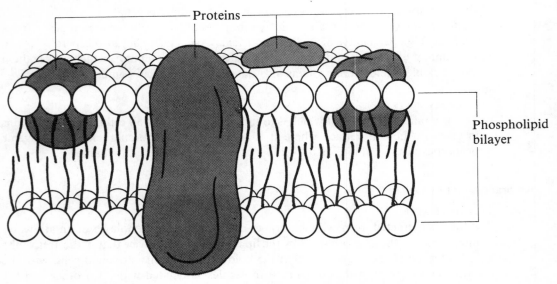

Phospholipid
bilayer

Figure 5.3   The fluid-mosaic model of membrane structure.

structure that is widely accepted today is that proposed by Singer and Nicolson in 1972 who suggested that proteins could be associated with a lipid bilayer in a number of ways (Figs. 5.2 and 5.3). The hydrophobic amino acid residues of the protein are in close contact with the hydrophobic side chains of phospholipids and the hydrophilic amino acid residues are at the surface in contact with the aqueous environment. Physical and chemical studies have shown that the oligosaccharide side chains of the glycoproteins and glycolipids are always present on the outer membrane surface and never on the inside of the cell.

LIPID FLUIDITY

One of the main ideas of the Singer and Nicolson model is that the lipid bilayer is fluid at physiological temperatures, so that the phospholipid molecules can move freely in the plane of the membrane. Movement of the phospholipids across the bilayer is rare. This, together with the distribution of the carbohydrate, shows that membranes are distinctly *asymmetric*.

5.2 Membrane transport: cellular imports and exports

Membranes perform a variety of important functions, some of which are highly specialized but their principal role is to control the flow of ions, metabolites and foreign compounds, such as drugs, into and out of the cell and between the various cellular compartments. This selective permeability leads to the concentration of ions and molecules and enables metabolism to take place in a controlled environment.

Transport is the movement of substances across membranes and this can occur by diffusion or by means of a carrier. These two categories of movement are known as *non-mediated transport* and *carrier-mediated transport*. Membrane transport can also be described as either *passive* or *active*. In the case of passive transport, molecules move down a concentration gradient whereas in active transport, the molecules move up a concentration gradient and require the expenditure of metabolic energy.

Non-mediated transport

RATE OF TRANSPORT

In this type of transport, substances move naturally from a higher to a lower concentration by *simple diffusion* and the *flux* or rate of transport is directly proportional to the concentration gradient across the membrane (Fig. 5.4).

Flux = amount of substance transported
      per unit area of membrane
      per unit time.
Flux = $\mu\text{mol cm}^{-2}\text{ s}^{-1}$

Examples of simple diffusion include the transport of oxygen and carbon dioxide in the lung and ethanol in the stomach.

MEMBRANE LIPIDS

The most important factor that affects the diffusion of a compound across membranes is its lipid solubility. As a general rule, the more hydrophobic and less polar a compound, the greater the rate of diffusion. The rate of transport also depends on the lipid composition of the membrane and particularly on the chain length and degree of saturation of the fatty acid side chains present in the phospholipids. Studies with model membranes such as liposomes have shown that the flux of a particular compound across a membrane falls with increasing chain length and also with the degree of saturation of the fatty acids. The rate of diffusion also falls when cholesterol is added to model membranes.

CHARGE

At neutral pH values, membranes carry a negative charge due to the polar head groups of the phospholipids or oligosaccharide chains on the surface and this charge modulates the membrane permeability by attracting cations and repelling anions. Ions, in general, do not readily cross membranes by simple diffusion, because of their charge and their high degree of hydration which makes them very hydrophilic.

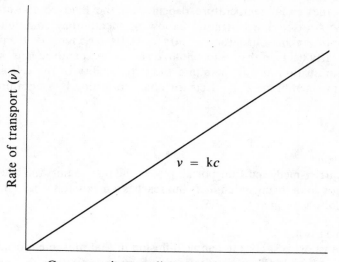

Concentration gradient across membrane (c)

Figure 5.4 The dependence of simple diffusion on the concentration gradient across the membrane.

TEMPERATURE

The permeability of membranes is dependent on the temperature. When membrane lipids are cooled sufficiently, a point is reached when they are no longer fluid and the membrane permeability is restricted. The temperature at which this occurs is known as the *thermotropic transition temperature*, when the packing of the membrane phospholipids changes from a mobile *liquid crystal* type of structure to a more rigid *crystal* state (Fig. 5.5). For many biological membranes, this is somewhere between 20

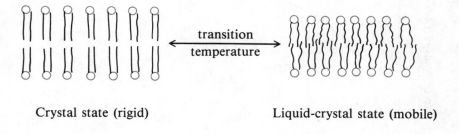

Crystal state (rigid)   Liquid-crystal state (mobile)

○ polar head groups

{ fatty acid side chains

Figure 5.5 The phase transition of phospholipids between the crystal and the liquid-crystal in membranes.

and 25 °C but the actual temperature depends on the lipid composition of the membrane. The transition temperature is low in membranes containing a high proportion of short-chain or unsaturated fatty acids. Bacteria increase the proportion of unsaturated fatty acids in their phospholipids as the temperature falls, in order to avoid a phase transition and to maintain the permeability of the cell membrane. Cholesterol in the membrane also affects the phase transition by smoothing out any drastic changes.

Carrier-mediated transport

RATE OF TRANSPORT
In the case of carrier-mediated transport, a plot of the flux against the concentration gradient does not go on rising indefinitely but reaches a maximum when all the carrier molecules are fully saturated (Fig. 5.6).

FACILITATED DIFFUSION
This type of transport is similar to simple diffusion in that movement occurs down a concentration gradient but the rate of transport is much faster than that expected for simple diffusion. The reason for this is that the molecules are transported by a carrier across the membrane and the diffusion constant for the carrier–molecule complex is greater than that for the molecule alone. Glucose and many amino acids are very hydrophilic molecules and so would not be expected to readily penetrate membranes by simple diffusion, but they do pass from the blood into cells quite freely by facilitated diffusion.

Facilitated diffusion also differs from simple diffusion in that it shows a high degree

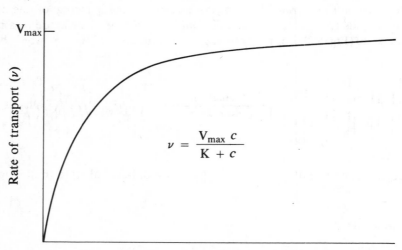

$$\nu = \frac{V_{max}\, c}{K + c}$$

Concentration gradient across membrane (c)

Figure 5.6 The dependence of carrier-mediated transport on the concentration gradient across the membrane.

of specificity due to the stereochemical configuration of the carrier. An example of this is seen in the transport of monosaccharides in the small intestine where D-glucose and D-galactose are rapidly transported whereas D-mannose and 2-deoxy-D-glucose are not. The structural requirements for sugar transport are a pyranose ring, a D configuration and a free —OH group in the 2-position. If the —OH groups in any other position are blocked, the sugar is still transported.

Facilitated diffusion can be inhibited by structurally related compounds which compete for the binding site on the carrier membrane-protein and an example of this is the inhibition of D-glucose transport by D-galactose.

D-glucose   D-galactose

D-mannose   2-deoxy-D-glucose

ACTIVE TRANSPORT

Active transport also involves a carrier and shows many of the features of facilitated diffusion including a high degree of specificity and saturation at high concentrations. The difference is that compounds are transported against a concentration gradient and require a source of metabolic energy which is usually provided by the direct or indirect hydrolysis of ATP.

Primary active transport　Active transport that uses the hydrolysis of ATP directly is known as primary active transport; probably the best known example of this is the process that maintains the low level of Na^+ in living cells compared to the extracellular fluid (Section 2.1). This low cellular Na^+ is sustained by pumping Na^+ out of the cell against a high concentration gradient, the energy being provided by the hydrolysis of ATP. The dependence of the Na^+ pump on ATP can be demonstrated by the addition of low concentrations (0.1 $\mu mol/litre$) of the cardiac glycoside oubain which inhibits the membrane ATPase and also the Na^+ pump. The result of this is that K^+ diffuse into the cells to replace the expelled Na^+ and to maintain electrical neutrality.

Secondary active transport   Some compounds are taken up by active transport systems that use the energy of a concentration gradient rather than the hydrolysis of ATP: this is known as secondary active transport. An example of this is the reabsorption of glucose in the kidney which is driven by the Na^+ gradient between the interior and the exterior of the renal tubules.

Transport of macromolecules

Large molecules are unable to cross membranes by the mechanisms described so far but macromolecules and particles can enter and leave cells by endocytosis and exocytosis.

ENDOCYTOSIS

In this process, the plasma membrane changes shape and fuses to form a vesicle which is then taken into the cell (Fig. 5.7a). Material that is outside the cell at this point or on the surface is therefore engulfed and encapsulated. This occurs when food is taken up by amoebae and when hormones such as insulin bind to their surface receptor and then become internalized.

EXOCYTOSIS

Macromolecules are able to leave the cell by exocytosis (Fig. 5.7b) which is essentially the reverse of endocytosis although the detailed mechanism is different. Examples of this can be seen in the pancreas which manufactures digestive enzymes (α-amylase and lipase) and hormones (insulin and glucagon) which are then exported from the tissue.

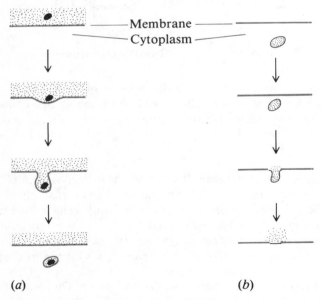

(a)   (b)

Figure 5.7   A schematic representation of endocytosis and exocytosis: (a) endocytosis; (b) exocytosis.

5.3   Prokaryotic cells: the simplest forms of life

Classification of cells

Cells are the basic units of all living systems and the higher animals and plants are *multicellular* as they are made up of millions of different types of cells. The simpler forms of life are *unicellular* and a particular organism consists of only one type of cell although there is considerable variety of shapes and sizes among the species. Examination of cells by light microscopy shows that almost all those from multicellular species and many unicellular organisms as well contain a clearly visible nucleus which is not present in cells from the very simplest forms of life. Cells can be divided into two distinct groups: the *eukaryotic cells* which contain a nucleus and the *prokaryotic cells* in which this structure is absent. These names are derived from the Greek words: *karyon* = kernel; *eu* = true; and *pro* = before. The structural feature that both prokaryotic and eukaryotic cells have in common is a membrane, about 8 nm thick, which completely encloses the cytoplasm and separates the cell from the external environment and its neighbours. In the case of eukaryotic organisms, membranes are also present in the cytoplasm surrounding the organelles and dividing the cell into different compartments. Cells and their components can therefore be considered as the next level of structural complexity in the organization of biomolecules beyond that of membranes.

Simple life forms

Prokaryotes are classified in the Kingdom Prokaryotae which is further divided into two groups: Division I are the cyanobacteria and Division II are the bacteria.

CYANOBACTERIA

These organisms, which are also known as *blue-green algae*, are able to carry out photosynthesis and also fix nitrogen from the atmosphere (Section 2.3). Their nutritional requirements are very basic as they need only a simple aqueous medium, exposure to light and a supply of carbon dioxide and nitrogen in order to grow. These tiny creatures play an important part in the biology of our planet since they appear to be responsible for all the nitrogen fixed in the oceans of the world.

MYCOPLASMAS

These organisms are one of the subdivisions of bacteria and are the simplest of the prokaryotic cells. They are different to other bacteria in that they do not possess a rigid cell wall and are very small with a diameter of 0.1 μm or less. Their flexible shape and small size makes them difficult to remove by filtration and this can be a problem when trying to prepare sterile solutions for cell culture work.

Most mycoplasmas live harmlessly as parasites in animals and plants but a few are pathogenic.

BACTERIA

This group of micro-organisms are the most abundant of the prokaryotes and are diverse in both structure and metabolism (Table 5.1). Some bacteria (*Rhizobium*) can fix nitrogen from the atmosphere while others (*Rhodopseudomonas*) have the ability to synthesize organic molecules from atmospheric carbon dioxide by photosynthesis. Most bacteria are harmless to man but some are pathogenic (Table 5.2) and, not surprisingly, this latter group have been extensively investigated.

Table 5.1 The diversity of bacteria

Structure			
Shape	*Description*	*Example*	*Size* (1 μm = ⊢—⊣)
Sphere	coccus	*Staphylococcus aureus*	◯
Rod	bacillus	*Bacillus anthracis*	⬭
Screw	spirillum	*Spirillum bdellovibrio*	〜〜
Curved rod	vibrio	*Vibrio cholera*	〜

Metabolism	
Description	*Nutritional requirement*
Aerobes	Oxygen
Anaerobes	Function without oxygen
Autotrophs	Water, inorganic salts and carbon dioxide
Heterotrophs	Water, salts and organic compounds

Table 5.2 Some examples of pathogenic bacteria

Bacteria	Disease
Staphylococcus aureus	Boils
Streptococcus pyogenes	Scarlet fever
Shigella dysenteriae	Dysentery
Salmonella typhi	Typhoid
Yersinia pestis	Plague
Clostridium tetani	Tetanus
Mycobacterium tuberculosis	Tuberculosis

Structure of prokaryotes

In spite of their wide diversity, there are a number of features that prokaryotic cells have in common. All prokaryotes have a relatively simple structure and the few details that are visible can only be seen with an electron microscope. They occur in a range of sizes from mycoplasmas (0.1 μm), to bacteria (1 μm) and blue-green algae (10 μm). The cytoplasm does not contain a nucleus or any other membrane-bound organelles and apart from the mycoplasmas, the cell membrane is surrounded by a rigid cell wall.

The bacterium *Escherichia coli* is an organism whose biochemistry has been extensively investigated and which shows many of the features of a 'typical' prokaryotic cell (Fig. 5.8).

CELL WALL

The osmotic pressure of the cell is greater than that of the environment and the porous but rigid cell wall serves to protect the cell against osmotic and mechanical damage.

The main structure of the cell wall is a peptidoglycan which is a gigantic molecule

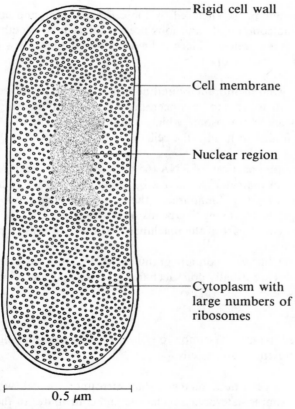

Rigid cell wall

Cell membrane

Nuclear region

Cytoplasm with large numbers of ribosomes

0.5 μm

Figure 5.8 The structure of a 'typical' prokaryotic cell.

that completely surrounds the bacteria like a sac (Section 4.2). The wall also contains other molecules depending on the particular species of bacteria. For example Gram-negative bacteria such as *E. coli* have an outer coat of *lipopolysaccharide* that covers the underlying peptidoglycan. Finally, the cell wall also governs the antigenic specificity of the organisms by means of short oligosaccharide chains present on the surface.

CELL MEMBRANE

Immediately beneath the cell wall is the cell membrane which acts as a selective permeability barrier allowing water, ions and nutrients to enter and excretory and waste products to leave the cell.

Particular regions of the cell membrane contain the components of the electron transport chain involved in the capture of metabolic energy as ATP. In the case of photosynthetic organisms, specialized areas of the membrane contain chlorophyll and other pigments needed for photosynthesis.

CYTOPLASM

This describes the aqueous phase enclosed by the cell membrane and contains the macromolecules, organic compounds and ions needed for cellular metabolism. The cytoplasm appears to be featureless under the light microscope but a few details can be seen with the electron microscope.

Nuclear region There is no membrane-bound nucleus as in eukaryotic cells but the genetic material of the cell appears to be concentrated in one particular region. In the case of *E. coli* there is one chromosome which is a single molecule of DNA about 1 mm long but highly folded to fit into the cell.

Ribosomes These are small particles of RNA (65 per cent) and protein (35 per cent) which are about 18 nm in diameter. An active bacterial cell may contain 10 000 or more of these structures which are found free in the cytoplasm or as *polysomes* bound to a thread of messenger RNA rather like beads on a string.

Ribosomes are an essential part of the machinery for the synthesis of proteins.

Storage granules Some bacteria contain granules which are food stores for the organism. These granules are usually polysaccharides such as glycogen but some are polymers of fatty acids.

OTHER STRUCTURES

The structures described so far are common to virtually all prokaryotic cells but other features may also be present in some cells.

Capsule This is a loose gelatinous material that surrounds the cell wall of some bacteria. It is usually a simple heteropolysaccharide and contributes to the antigenic properties of the cell.

Streptococci have a small capsule of hyaluronic acid while the tubercle bacillus has a waxy coat which gives considerable protection to the organism against physical and chemical damage.

Flagella   These structures, where they are present, are responsible for the motility of the cell and consist mainly of the protein flagellin.

Pilli   In some cases the surface is covered with thread-like tentacles which are probably used for adherence to other structures.

Antibiotics

The differences in structure of prokaryotes and eukaryotes are quite marked but the metabolic pathways of the simplest organisms are very similar to those in man. There are however small but subtle differences in the metabolism of the two classes of organisms and these have been exploited in developing drugs that are toxic to invading bacteria but which are harmless to the infected organism. These drugs, which are natural compounds produced largely by moulds, are known as *antibiotics* (*anti* = against; *bios* = life).

PENICILLIN

This was the first antibiotic discovered and acts by preventing the incorporation of D-alanine into the peptidoglycan of the bacterial cell wall. Animals are unaffected as they do not possess a rigid cell wall and use L and not D amino acids in their metabolism.

ERYTHROMYCIN

This compound binds to the 50 S subunits of bacterial ribosomes and blocks the translocation reaction of protein synthesis. Erythromycin does not bind to the equivalent subunit found in animals which is slightly larger (60 S) so that mammalian protein synthesis is unaffected.

STREPTOMYCIN

This antibiotic interacts with the 30 S subunit of bacterial ribosomes and interferes with the translation part of bacterial protein synthesis. Streptomycin does not bind to the slightly larger ribosomal subunit (40 S) found in animals so that mammalian protein synthesis is unaffected.

TETRACYCLINES

These antibiotics, which affect a large number of bacteria, also bind to the 30 S subunit of bacterial ribosomes and prevent the elongation of the peptide chain. Tetracyclines also interact with the 40 S subunit of eukaryotic ribosomes *in vitro* but have no effect on eukaryotic protein synthesis *in vivo* as they do not penetrate eukaryotic membranes.

5.4 Eukaryotic cells: the units of higher life forms

Size

VOLUME

Eukaryotes are bigger than prokaryotes and most animal and plant cells have a diameter at least 10 times that of bacteria. This size difference is even more marked if volumes are compared, when eukaryotic cells are seen to be 1000 to 10 000 times bigger than a typical prokaryotic organism such as a streptococcus (Table 5.3). The much greater volume of eukaryotic cells means that the distances molecules can travel within the cytoplasm are very large, so that the chances of random collisions between a given pair of molecules will be very small when compared to prokaryotic cells. Chemical reactions in the cell depend on such collisions taking place, so that a large

Table 5.3 The relative sizes of prokaryotic and eukaryotic cells

Cell	Diameter (μm)	Approximate volume (μm^3)
Prokaryotic cells		
Mycoplasma	0.1	0.0005
Streptococcus	1	0.52
Eukaryotic cells		
Yeast	10	520
Rat liver	22	5 600
Corn leaf	24	7 200
Human ovum	120	905 000

increase in volume would appear to reduce the efficiency of metabolism and perhaps even render it impossible. The problem arising from this marked increase in size can be overcome if the cell is divided up into a series of different compartments so that the volume actually available for diffusion is restricted. This compartmentalization is achieved by a series of membranes which enclose and separate different areas of the cell so that particular metabolic sequences can be localized and thereby controlled. Biochemists have been able to isolate the different parts of eukaryotic cells and to study their structure and metabolism by disrupting the cells in a buffered isotonic medium and separating their components by centrifugation.

Diversity and complexity

TISSUES

Some eukaryotes such as yeast are unicellular organisms but most of them are multicellular animals and plants that contain vast numbers of different cells. In order to cope with the complexity of higher organisms, groups of cells are organized together in *tissues* whcih are a collection of cells that perform a similar function.

Examples of tissues are bone, muscle and nervous tissue in animals and parenchyma, phloem and supporting tissue in plants.

ORGANS

Some tissues operate by themselves but more often than not several tissues are arranged together to form an *organ* which performs a particular function. In animals, for example, epithelial tissue, muscle, blood vessels and nerves are organized together to form the stomach where the initial digestion of food takes place. Similarly, in plants, several tissues combine to form leaves whose primary function is photosynthesis.

NUTRITION

The increased complexity of eukaryotic cells means that they are nutritionally more demanding than prokaryotes and require a supply of organic compounds as food, and not just simple inorganic materials, in order to grow and divide. Eukaryotic cells vary widely in structure and function, depending on their physiological role in the organism, and occur in a variety of shapes and sizes. However, in spite of this diversity, there are a number of structural and metabolic features that are common to almost all cells and the next section (Section 5.5) deals with the distribution of metabolic function in terms of the structures found in 'typical cells' from the animal and plant kingdoms (Figs. 5.9 and 5.10).

Plant cell walls

Plant cells are surrounded by a rigid cell wall from 0.1 to 10 μm thick which is made up of several polysaccharides and a small amount of glycoprotein. The major component is cellulose and this is present as fibres set in a polysaccharide matrix rather like steel rods in reinforced concrete. Such an arrangement gives considerable rigidity so that the cell wall is the skeleton of the plant and controls its shape and size. The cell wall also acts like a skin in that it protects delicate internal tissues from mechanical injury.

Eukaryotic membranes

Eukaryotes have a more complex structure than prokaryotes and electron microscopy shows that they contain a number of different membranes and membrane-bound structures.

PLASMA MEMBRANE

The plasma membrane of eukaryotes is about 7 nm thick and performs the same role as the cell membrane of prokaryotes, namely the control of the passage of materials into and out of the cell. In some cases the membrane is convoluted with outward folds known as *microvilli*. These folds greatly increase the surface area of the membrane and microvilli are present in cells where the uptake (kidney) and excretion (pancreas) of molecules is important. The plasma membrane can also have inward folds and some invaginations lead into the depth of the cell to form a complex 3-D network known as the endoplasmic reticulum (ER).

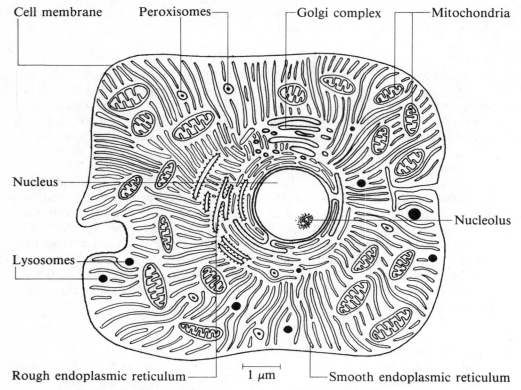

Figure 5.9   The general structure of a 'typical' animal cell.

ENDOPLASMIC RETICULUM

The endoplasmic reticulum is an interconnected series of membranes with flattened areas that form *cisternae* and isolated portions known as *vesicles*. It permeates the cytoplasm and is connected with the plasma and nuclear membranes. The endoplasmic reticulum is present in most animal cells although its extent varies with the type of cell. Two forms are visible under the electron microscope, the so-called rough and the smooth endoplasmic reticulum.

Rough endoplasmic reticulum   This consists of cisternae studded with ribosomes which give the characteristic rough appearance to the membranes. The rough ER is the site of protein synthesis and this structure is quite extensive in organs such as the liver and pancreas that are active in the synthesis of proteins.

Smooth endoplasmic reticulum   The membranes of the smooth ER tend to be tubular rather than flattened and have a smooth appearance due to the absence of ribosomes. The smooth ER therefore does not synthesize proteins but it is involved in their modification by adding short chains of oligosaccharide. The membranes also contain

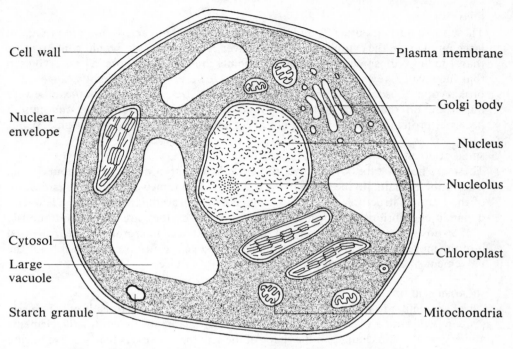

Figure 5.10 Schematic representation of a 'typical' plant cell.

cytochrome P-450, which is needed for the oxidation of foreign organic compounds such as drugs, pesticides and toxins.

GOLGI APPARATUS

This membranous structure is a specialized region of the endoplasmic reticulum and consists of flattened and slightly curved membranes with a disc shape. There are also granules associated with these membranes which are vesicles packed with material synthesized at this site. The Golgi region can be likened to a packing department of a factory for it is here that macromolecules are modified by glycosylation and sulphation, sorted then packaged ready for export from the cell.

Cytoplasm

The cytoplasm accounts for about half of the volume of the cell and contains the soluble enzymes and metabolites.

STORAGE GRANULES AND DROPLETS

The cytoplasm of some cells contains stores of food. These can take the form of granules of polysaccharides such as starch in plants and glycogen in liver and muscle or droplets of fat such as triglyceride in adipose tissue and the oil-rich seeds of some plants.

RIBOSOMES

These small particles of RNA and protein are about 30 nm across and are present in large numbers in the cytoplasm. A typical liver cell, for example, would contain more than 1 million ribosomes. Eukaryotic ribosomes are similar to those in prokaryotes in that they play a key role in protein synthesis. However, they are slightly bigger than prokaryotic ribosomes and this small difference in size and composition can be very important, in that some antibiotics bind to prokaryotic ribosomes and inhibit bacterial protein synthesis but do not affect eukaryotic ribosomes (Section 5.3).

MICROTUBULES

These are hollow tubes of the glycoprotein *tubulin* with a diameter of about 25 nm. Microtubules maintain the shape of the cell, provide a framework for the organization of the internal structures and act as channels for intracellular transport. There is a dynamic equilibrium of the internal microtubules as they are being continuously broken down and formed. In contrast to this, the external microtubules are organized into the more permanent structures of *flagella* and *cilia* which are responsible for the movement of the cell and of materials close to its surface.

MICROFILAMENTS

These are similar to microtubules but smaller and consist of fine threads of the muscle protein *actin* with a diameter of about 7 nm. Microfilaments are contractile elements and are involved in changes in the shape of the cell: they are particularly active during endocytosis and exocytosis when the plasma membrane changes shape to engulf or extrude material.

Organelles

The most striking structural difference between prokaryotes and eukaryotes is the presence in eukaryotic cells of a number of membrane-bound organelles and the structure and metabolism associated with these bodies is the subject of Section 5.5.

5.5 Cellular organelles: the localization of metabolism

Nucleus

The most striking feature of eukaryotic cells is the presence of a nucleus (Fig. 5.11a). This is the largest organelle in the cell and because of its size (3–10 μm) it is clearly visible by light microscopy. It is a constant and essential component of cells and those few that lack a nucleus, such as erythrocytes and the lens of the eye, are incapable of growth and division. The nucleus is bounded by the *nuclear envelope* which is a double membrane with pores for transport and encloses the *nucleoplasm*.

NUCLEOPLASM

The nucleus contains 99 per cent of the cellular DNA which is associated with basic proteins called *histones* and RNA to form *chromatin*, the genetic material of the cell. Chromatin is normally distributed throughout the nucleoplasm but when the cell starts to divide, the chromatin becomes organized into distinct linear structures known as *chromosomes* which are then duplicated in the daughter cells by replication of DNA. The nucleus also controls the growth of the cell by synthesizing the RNA needed for protein synthesis.

NUCLEOLI

The nucleus contains one or more *nucleoli* which are bodies, particularly rich in RNA, with a diameter in the region of 1 μm. They are an area of the nucleus that is responsible for the production and storage of RNA needed for the assembly of ribosomes.

Mitochondria

These particles can just about be seen under the light microscope when stained with a suitable dye but the details of their structure are only visible under the electron microscope. The organelles are surrounded by a double membrane and the inner membrane is highly folded to form *cristae* (Fig. 5.11b). The organelles are cylindrical or spherical in shape and their size and number vary with the type of cell examined. Yeast, for example, contains a single giant mitochondrion while liver has more than 1000 mitochondria which occupy 22 per cent of the cell volume. Mitochondria have been accurately described as the 'power plants' of the cell as they are the site of several oxidative processes and it is here that the oxidation of fats and carbohydrates takes place with the production of carbon dioxide and water. The *matrix* contains the enzymes of fatty acid oxidation and the tricarboxylic acid cycle (Krebs' cycle) where the carbon dioxide is produced; the *inner membrane* is where the respiratory chain is located which gives rise to the product, water. Most of the metabolic energy is released in the respiratory chain where it is 'captured' as ATP.

Chloroplasts

These organelles, which are unique to plants, are larger than mitochondria and have a length of about 5 μm. There are structural similarities to mitochondria in that the

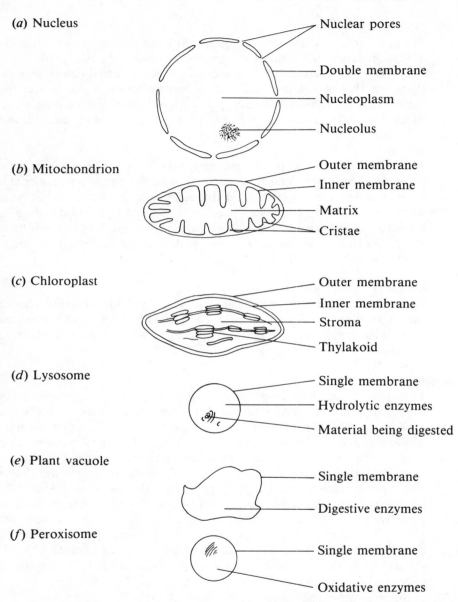

(a) Nucleus

Nuclear pores

Double membrane

Nucleoplasm

Nucleolus

(b) Mitochondrion

Outer membrane

Inner membrane

Matrix

Cristae

(c) Chloroplast

Outer membrane

Inner membrane

Stroma

Thylakoid

(d) Lysosome

Single membrane

Hydrolytic enzymes

Material being digested

(e) Plant vacuole

Single membrane

Digestive enzymes

(f) Peroxisome

Single membrane

Oxidative enzymes

Figure 5.11   Some common organelles found in animals and plants. The cross-sections of these organelles are not all drawn to the same scale. Details of their relative sizes are shown in Fig. 5.10 and are given in the text.

chloroplasts of higher plants are surrounded by a double membrane and have an internal system of membranes with a highly ordered structure (Fig. 5.11c). There is also a functional similarity to mitochondria in that both organelles are concerned with energy transfer. In mitochondria, organic compounds are oxidized to carbon dioxide and water with the release of energy whereas in chloroplasts, the energy of sunlight is captured and used to synthesize organic compounds from carbon dioxide and water. The carbohydrates produced during photosynthesis can then be stored as a polymer, such as starch, or transported as sucrose for metabolism in other parts of the plant.

Lysosomes

These are spherical organelles with a diameter in the region of 0.2 to 0.5 μm: they are bounded by a single membrane (Fig. 5.11d). They are quite numerous and a liver cell contains about 300 of these organelles. Lysosomes have no internal structure and contain a range of acid hydrolases capable of breaking down the molecules and macromolecules of the cell. The organelles can be thought of as scavengers since they are involved in the digestion of exogenous materials taken into the cell, such as food particles in the amoeba, and endogenous compounds and structures during the normal turnover of cell components.

Lysosomes also participate in other important physiological processes where the controlled digestion of cellular material is important. This includes the penetration of the ovum by the spermatozoa during fertilization, the involution of the uterus following birth and the metamorphosis of amphibian larvae.

Plant vacuoles

One of the features of plant cells is the presence of one or more vacuoles which are large structures surrounded by a single membrane. They contain a high concentration of molecules and ions so that water readily enters by osmosis and they swell and can occupy quite a large part of the cell (Fig. 5.11e). These organelles are similar to lysosomes in that they contain hydrolytic enzymes and these are active during intracellular digestion and the shedding of leaves and flowers. Vacuoles are also a store for nutrients and specialized products such as toxic alkaloids, which protect the plant, and anthocyanins, which are responsible for the bright colours of the leaves and flowers.

Peroxisomes

These particles are the same size and occur with about the same frequency as lysosomes: they are found in a number of tissues including liver, kidney and the leaves of plants. They are bounded by a single membrane and contain crystals of a limited number of oxidative enzymes which produce hydrogen peroxide as a by-product (Fig. 5.11f). This is potentially lethal to the cell but fortunately peroxisomes also contain the enzymes catalase and peroxidase so that the hydrogen peroxide is rapidly inactivated. Peroxisomes are also known as *microbodies*, a general term that includes *glyoxysomes* a specialized form of peroxisomes found in oil seeds. These plant organelles contain the enzymes of the glyoxylate cycle which allows the seeds to convert fat stored as triglyceride into carbohydrates needed for growth. Such a conversion of fat to carbohydrate is not possible in animals.

The division of cellular metabolism

The living cell is a highly organized and integrated structure which in many ways is similar to that of a factory. This analogy is used in Table 5.4 which summarizes the findings of this and the previous two sections by describing the main metabolic activities associated with the structures found in eukaryotic cells.

Table 5.4 The localization of metabolic activity in structures found in eukaryotic cells

Structure	[Factory analogy]
Plasma membrane and cell wall	[The wall and gates]
These boundaries protect and control the passage of materials into and out of the cell.	
Microtubules and microfilaments	[The building framework]
These structures provide a framework for the organization and shape of the cell.	
Nucleus	[The planning department]
This organelle, packed with genetic information, determines the growth and division of the cell and controls protein synthesis.	
Chloroplasts	[Electrical transformer]
These organelles, found only in plants, collect solar energy and use it in the synthesis of complex structures.	
Mitochondria	[The boiler and generator]
These 'power plants' of the cell oxidize foods and capture the free energy liberated as ATP for use in the cell.	
Storage granules and droplets	[Coal bunkers and oil tanks]
These are the food stores of the cell in the form of granules of glycogen or droplets of triglyceride.	
Rough endoplasmic reticulum	[The production line]
It is here that proteins are synthesized both for use in the cell and also for export.	
Golgi apparatus	[Packing and transport]
Proteins are modified in this structure by glycosylation and sulphation then packaged ready for export.	
Plant vacuoles	[The warehouse]
These organelles store specialized plant products such as alkaloids and plant pigments.	
Peroxisomes	[Ventilators for toxic wastes]
The oxidative enzymes in these organelles produce hydrogen peroxide as a toxic by-product then inactivate it with catalase.	
Smooth endoplasmic reticulum	[Incinerators for waste]
This part of the ER metabolizes potentially toxic foreign organic compounds by oxidation and renders them inactive.	
Lysosomes	[Cleaning department]
These organelles are concerned with the controlled digestion of intra- and extra-cellular molecules and debris.	

6. Enzymes

6.1 Enzymes: the biological catalysts

Catalysis

Metabolism takes place very rapidly in living organisms with many individual reactions complete in milliseconds or less. This fast rate of metabolism is only possible because of the enormous increase in the rate of chemical reactions brought about by biological catalysts known as *enzymes*. An understanding of metabolism is therefore only possible by being aware of the properties of these ubiquitous catalysts.

1. Enzymes do not initiate new reactions.
2. They speed up the rate of a reaction.
3. They have immense catalytic activity.
4. They are highly specific.

CATALASE

These points can be illustrated by reference to the enzyme catalase which catalyses the breakdown of an aqueous solution of hydrogen peroxide to water and oxygen.

$$2\,H_2O_2 \longrightarrow 2\,H_2O + O_2$$

1. This reaction takes place spontaneously but at a very low rate to give a barely detectable release of oxygen.
2. The addition of catalase accelerates the rate of breakdown of the hydrogen peroxide by about 10^6 to give a vigorous evolution of oxygen.
3. One molecule of catalase, which contains four active sites, catalyses the breakdown of about 90 000 molecules of hydrogen peroxide per second.
4. Catalase is very specific and does not act on other substrates.

The catalytic power of enzymes

Enzymes in living organisms catalyse reactions under relatively mild conditions, whereas the chemist working in the laboratory frequently has to use strong acid or alkali and powerful oxidizing and reducing agents, often at an elevated temperature. The catalytic power of enzymes can be seen, for example, in the case of the digestive proteases which catalyse the hydrolysis of proteins in the gut during the digestion of food. The mild conditions under which they operate are in marked contrast to those required to carry out the reactions in the laboratory (Table 6.1).

Table 6.1   The breakdown of proteins to amino acids

Conditions	Chemical hydrolysis	Enzymatic hydrolysis
Place	Laboratory	The gut
Temperature	110 °C	37 °C
pH	6 M HCl (very acid)	Most at pH 5–6 (slightly acid)
Pressure	High (sealed tube)	One atmosphere (open tube)
Time	18 h	1–3 h
Loss of amino acids	Partial or complete destruction of some	None lost

Equilibrium position

Unlike the breakdown of hydrogen peroxide, many of the chemical reactions in living matter are reversible and in this case enzymes speed up the rate of the reaction in both directions but do not affect the final position of equilibrium. For example, the final equilibrium mixture of acid, alcohol and ester is the same for the reaction below whether the catalyst is acid (H^+), alkali (OH^-) or the enzyme liver esterase.

$$CH_3COOH + C_2H_5OH \rightleftharpoons CH_3COOC_2H_5 + H_2O$$

The energy of activation

The progress of a chemical reaction can be illustrated by the type of diagram shown in Fig. 6.1, in which the reactant is shown on the left and the products on the right of a barrier like a hill. Using this physical analogy, it is clear that the products are at a lower energy than the reactants but the reaction does not proceed spontaneously because of the barrier posed by the *energy of activation*. One way to start the reaction is to increase the energy of the molecules by applying heat so that they have sufficient

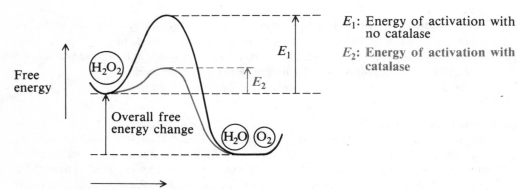

E_1: Energy of activation with no catalase

E_2: Energy of activation with catalase

Figure 6.1   Energy diagram of the breakdown of hydrogen peroxide in the absence (black) and the presence (red) of the enzyme catalase.

kinetic energy to overcome the barrier. This is the method commonly used in the laboratory but it is not possible in living matter which operates at a low and often constant temperature. The alternative way is to channel the reaction through a different pathway with a lower energy of activation, so that enough molecules possess sufficient kinetic energy to cross the barrier even at physiological temperatures and this is basically how enzymes operate.

The active site

This lowering of the energy barrier is brought about by the three-dimensional shape of the enzyme molecule which gives rise to the region known as the *active site*. This part of the enzyme protein binds the reacting molecules so that they are held in exactly the right position in space to react. Such an arrangement is considerably better than the random collisions which occur in free solution as only rarely would the molecules come together with the correct orientation to react. The amino acid residues at the active site also play an active role by donating or withdrawing electrons from the functional groups on the substrate. The forces that bind the substrate at the active site are relatively weak so that the products can be released from the enzyme surface when the reaction is complete (Fig. 6.2).

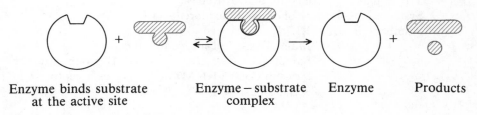

Enzyme binds substrate   Enzyme – substrate   Enzyme   Products
at the active site   complex

Figure 6.2 The active site of an enzyme.

The active site is only a small part of the enzyme protein and parts of the molecule away from the active site can sometimes be removed without affecting the activity. However, this is not usually possible since the three-dimensional shape taken up by the protein depends on the composition of all the amino acids and removal of part of the polypeptide chain normally causes a change in shape of the active site and subsequent loss of activity. The rest of the molecule, away from the active site, is therefore necessary to allow a folding of the protein which brings together in space amino acid residues that may be far apart along the polypeptide chain.

The active site is often in a fold or cleft in the enzyme into which the substrate molecules fit. This led Fischer to suggest the *lock-key hypothesis* in which the close binding of the substrate to the active site is likened to that of a key fitting in a lock. This idea, although essentially correct, gives a rather static picture of the active site which actually changes shape to accommodate the substrate as it approaches. This modification proposed by Koshland is known as the *induced-fit hypothesis*.

Table 6.2 The international classification of enzymes

Group No.	Class	Reaction catalysed
1	Oxidoreductases	Oxidation/reduction reactions.
2	Transferases	The transfer of functional groups.
3	Hydrolases	Hydrolytic reactions.
4	Lyases	The addition of groups to double bonds and the reverse reaction.
5	Isomerases	The rearrangement of molecules to give isomers.
6	Ligases	The formation of bonds coupled to the cleavage of ATP.

Enzyme classification

Enzymes are classified into six groups depending on the reaction catalysed (Table 6.2). Complete identification of an enzyme is possible from its systematic name and number. The first number denotes the main class, the second the sub-class, the third the sub-sub-class and the fourth is the individual number of the enzyme within the class. An example showing how this classification operates is the enzyme alcohol dehydrogenase which catalyses the oxidation of ethanol to acetaldehyde.

$$C_2H_5OH + NAD^+ \rightleftharpoons CH_3CHO + NADH + H^+.$$

Trivial name:	alcohol dehydrogenase
Systematic name:	alcohol: NAD oxidoreductase
Number:	1.1.1.1.

	Main class:	1 an oxidoreductase
	Sub-class:	1 acts on —CHOH group of donor
	Sub-sub-class:	1 NAD or NADP as acceptor
	Serial No.:	1

6.2 The activity of enzymes: substrate concentration and K_m

The measurement of enzyme activity

Enzymes are difficult to assay directly because of the very low concentrations that are present in living organisms. Their presence is therefore detected and measured by looking at the effect they have on their substrates. This is carried out by following the disappearance of the substrate or the appearance of the product with time.

PROGRESS CURVE

If a plot is made of the amount of substrate changed with time, a curve of the type shown in Fig. 6.3 is obtained: this is known as a *progress curve* as it shows the progress of the enzyme-catalysed reaction. At first there is a rapid increase in the amount of substrate changed but this soon declines as equilibrium is approached. The decline is

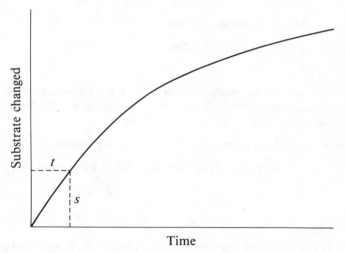

Figure 6.3 The progress curve for an enzyme-catalysed reaction. The enzyme activity (v) is the initial reaction rate taken from the linear part of the curve so that: $v = s/t$.

also due to the fall in substrate concentration as the reaction proceeds. In addition to this, the enzyme may be inhibited by the products of the reaction or become inactivated under the conditions of the experiment. For these reasons it is not usually possible to derive an equation to fit the curve. The *enzyme activity* therefore is taken from the initial or linear part of the curve, when the adverse effects are minimal and the activity is defined as the *amount of substrate changed per unit time.*

Enzyme activity = Initial reaction velocity

$$v = s/t$$

ENZYME UNITS

Enzyme activities are commonly given in terms of *international units* (U) such that:

$$1 \text{ U} = 1 \text{ } \mu\text{mol substrate min}^{-1}$$

The SI unit recommended by the IUB is the *katal* so that:

$$1 \text{ katal} = 1 \text{ mol substrate s}^{-1}$$

This unit is rather big and unwieldy and activities are more often expressed in *microkatals* (μkat) or *nanokatals* (nkat).

CATALYTIC-CENTRE ACTIVITY

The activity of enzymes can be compared by referring to their catalytic-centre activities. This is defined as the number of molecules of substrate transformed per second by a single catalytic centre.

$$\text{Catalytic-centre activity} = \text{katal} \times MW/n$$

$$MW = \text{molecular weight}$$

$$n = \text{number of active sites per molecule}$$

SPECIFIC ACTIVITY

The catalytic-centre activity can only be measured if the enzyme is pure and most laboratory preparations are impure. The activity is therefore usually expressed as units per unit weight of protein.

$$\text{Specific activity (international units)} = \text{U mg protein}^{-1}$$

$$\text{Specific activity (katals)} = \text{katal kg protein}^{-1}$$

Substrate concentration

PLOT OF v AGAINST s

The activity of an enzyme (v) depends on the concentration of substrate (s) and when v is plotted against s most enzymes give a curve similar to that shown (Fig. 6.4). The curve is a rectangular hyperbola and can be described by the equation:

$$(V - v)(K_m + s) = VK_m$$

where V and K_m are constants.

THE MICHAELIS EQUATION

Michaelis and Menten derived this equation from first principles by assuming that the enzyme (E) and the substrate (S) reacted to give an enzyme–substrate complex (ES) which then broke down to form the enzyme (E) and products (P):

$$E + S \underset{k_2}{\overset{k_1}{\rightleftharpoons}} ES \xrightarrow{k_3} E + P$$

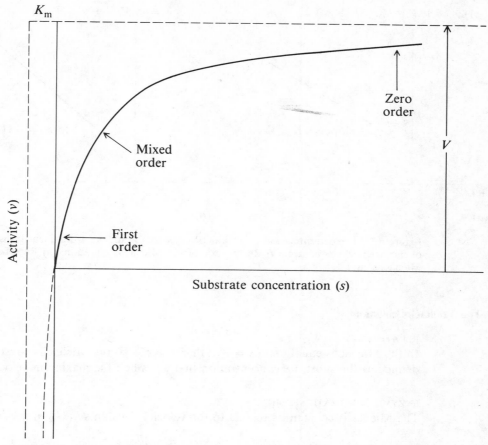

Figure 6.4 The variation of enzyme activity (v) with substrate concentration (s).

The more usual forms of the equations are:

$$v = V/(1 + K_m/s) \quad \text{or} \quad v = Vs/(s + K_m)$$

where V is the maximum possible activity of the enzyme when it is fully saturated with substrate and K_m is the Michaelis constant.

DETERMINATION OF K_m AND V

The kinetic constants can most easily be determined from a linear transformation of the Michaelis equation. This can be obtained by plotting $1/v$ against $1/s$ and determining $1/K_m$ and $1/V$ from the intercepts on the axis (Fig. 6.5a).

Multiplication of the reciprocal equation with s gives a plot of s/v against s which is better as it gives greater weight to the more accurate high substrate points (Fig. 6.5b).

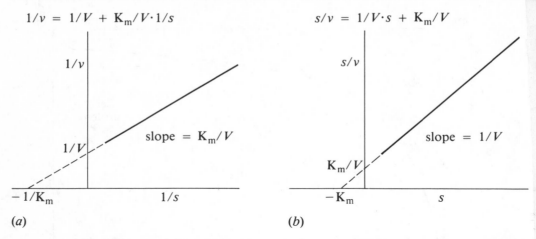

Figure 6.5 The determination of the kinetic constants V and K_m from linear transformations of the Michaelis equation. (*a*) Reciprocals of the Michaelis equation. (*b*) Reciprocals of the Michaelis equation $\times s$.

The Michaelis constant

RELATIONSHIP TO V

In the Michaelis equation, if $s = K_m$, then $v = V/2$ so the Michaelis constant can be defined as the substrate concentration that gives half the maximum velocity (V).

ENZYME–SUBSTRATE AFFINITY

The Michaelis constant is related to the velocity constants k_1, k_2 and k_3 so that:

$$K_m = (k_2 + k_3)/k_1$$

If k_3 is small compared to k_1 and k_2 then K_m approximates to k_2/k_1 which is the dissociation constant of the enzyme–substrate complex so that:

High K_m = high ES dissociation = low ES affinity

Low K_m = low ES dissociation = high ES affinity

The rate constant cannot be totally ignored so that the Michaelis constant is not usually the dissociation constant of the ES complex but it does give a measure of the enzyme–substrate affinity.

SUBSTRATE CONCENTRATION AND K_m

If the substrate concentration is small compared to the K_m then the Michaelis equation approximates to $v = Vs/K_m$. Under these conditions, the reaction velocity varies linearly with substrate concentration and the reaction is *first order*.

If s is large compared to the K_m, then the equation approximates to $v = V$. In this case the velocity is independent of substrate concentration so the reaction is *zero order*.

Concentrations of substrate between these two extremes give mixed order kinetics (Fig. 6.4).

K_m AND GLUCOSE METABOLISM

The K_m is not just a physical constant but is important in metabolism. This can be seen in the case of *hexokinase* and *glucokinase* which catalyse the phosphorylation of glucose by ATP to give glucose-6-phosphate.

$$\text{glucose} + \text{ATP} \xrightarrow{\text{Mg}^{2+}} \text{glucose-6-phosphate} + \text{ADP}$$

Skeletal muscle   The dominant enzyme in muscle is hexokinase which has a K_m of 0.1 mmol/litre glucose and this is considerably less than the level of glucose in the blood, which varies in health from 4 mmol/litre to 8 mmol/litre. The enzyme is therefore well saturated whatever the concentration of the blood glucose and this is very convenient for muscle which uses glucose even at low concentrations of blood glucose.

Liver   The dominant enzyme in liver is glucokinase which has a K_m of 10 mmol/litre, about 100 times greater than hexokinase. In the case of the liver enzyme the K_m is only slightly greater than the normal range of blood glucose concentrations so that glucokinase is very sensitive to changes in the concentration of glucose in the blood. A rise of glucose following a meal causes an increase in the activity of the enzyme and leads to an increased concentration of glucose-6-phosphate, which is converted to glycogen and stored for future energy requirements. A fall in blood glucose has the opposite effect so that less glucose is phosphorylated by the liver and glucose is available for other tissues.

6.3 Inhibitors of catalysis: kinetics and examples

The importance of enzyme inhibition

METABOLISM

There are many natural compounds that inhibit enzymes and changes in their cellular concentration reduce the enzyme activity and serve as a means of metabolic control. This aspect of enzyme inhibition is dealt with later under the general heading of metabolic control.

Natural and synthetic reagents have been used by biochemists to unravel the metabolic pathways of intermediary metabolism. Inhibition of a particular enzyme in a metabolic pathway leads to an accumulation of the substrate which can then be isolated and identified. Krebs used this approach when carrying out his classical work on the tricarboxylic acid cycle. He showed that the respiration of tissues was strongly inhibited by low concentrations (10 mmol/litre) of malonate which blocked the oxidation of succinate to fumarate thereby causing the accumulation of succinate.

$$
\begin{array}{ccc}
\text{CH}_2\text{COO}^- & & \text{CHCOO}^- \\
| & + \text{FAD} & \| & + \text{FADH}_2 \\
\text{CH}_2\text{COO}^- & & \text{CHCOO}^-
\end{array}
$$

Substrates: succinate   fumarate

Inhibitor: malonate

$$
\text{CH}_2 \big\langle {}^{\text{COO}^-}_{\text{COO}^-}
$$

APPLICATIONS

Enzyme inhibitors can also be of benefit to mankind as drugs, herbicides and pesticides. In these cases, the compounds are designed to inhibit a specific enzyme in the micro-organisms, plant or insect without affecting other plants or animals. Regrettably, enzyme inhibition has also been used for destructive purposes and cyanide used by murderers, both real and fictional, kills by inhibiting cytochrome oxidase. This particular enzyme, which is present in all aerobic cells, is essential for intracellular respiration and therefore the maintenance of life itself.

On a larger scale of murder, many of the toxic gases used in warfare act by inhibiting key enzymes.

Irreversible and reversible inhibitors

IRREVERSIBLE

Some inhibitors become covalently linked to the enzyme and are bound so strongly that they cannot be removed. In these cases the enzyme activity decreases with time

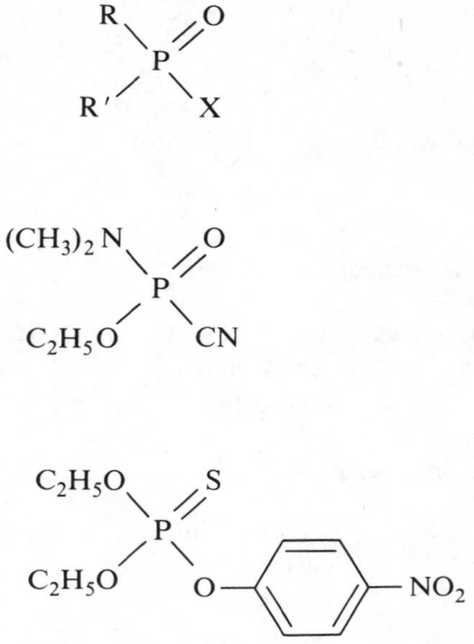

General formula

R, R′ = alkyl groups

X = — F, — CN

Tabun

A highly toxic nerve gas

Parathion

An insecticide, much less toxic to man but kills insects as they convert $P = S$ to toxic $P = O$ compounds.

Figure 6.6 Some examples of organophosphorus compounds.

and eventually reaches zero. The degree of inhibition is therefore defined by a velocity constant k_1:

$$E + I \xrightarrow{k_1} EI$$

This type of inhibition has disastrous consequences for metabolism and irreversible inhibitors are invariably toxic to living organisms.

Examples of this type of inhibitor are the *organophosphorus compounds* (Fig. 6.6) which are inhibitors of *acetylcholinesterase*, a key enzyme involved in the transmission of the nerve impulse at the neuromuscular junction and parts of the central nervous system.

Normally the enzyme reacts with the natural substrate acetylcholine to release the alcohol choline and form an acetylated complex which is then hydrolysed to give acetate and the free enzyme. In the presence of an organophosphorus compound, the first stage takes place but the phosphorylated intermediate is very stable and is not hydrolysed further (Fig. 6.7). There are a large number of these organophosphorus compounds, some of which are insecticides while others are nerve gases used in warfare (Fig. 6.6).

REVERSIBLE

In the presence of a reversible inhibitor, there is a rapid fall in enzyme activity with time to a reduced but constant level when equilibrium is reached. Reversible inhibition is therefore defined by an equilibrium constant K_1:

$$E + I \xrightleftharpoons{K_1} EI$$

Many inhibitors of interest to biochemists fall into this category.

Normal

(i) Formation of acetylated enzyme intermediate

$$\underset{\text{acetylcholine}}{CH_3CO \cdot O(CH_2)_2\overset{+}{N}(CH_3)_3} + \underset{\text{enzyme}}{E-OH} \rightleftharpoons \underset{\substack{\text{acetylated} \\ \text{enzyme}}}{CH_3CO-O-E} + \underset{\text{choline}}{HO(CH_2)_2\overset{+}{N}(CH_3)_3}$$

(ii) Hydrolysis of acetylated enzyme intermediate

$$\underset{\substack{\text{acetylated} \\ \text{enzyme}}}{CH_3CO-O-E} + H_2O \longrightarrow \underset{\text{acetate}}{CH_3COO^-} + H^+ + \underset{\text{enzyme}}{HO-E}$$

Inhibited by organophosphorus compound

(i) Formation of phosphorylated enzyme intermediate

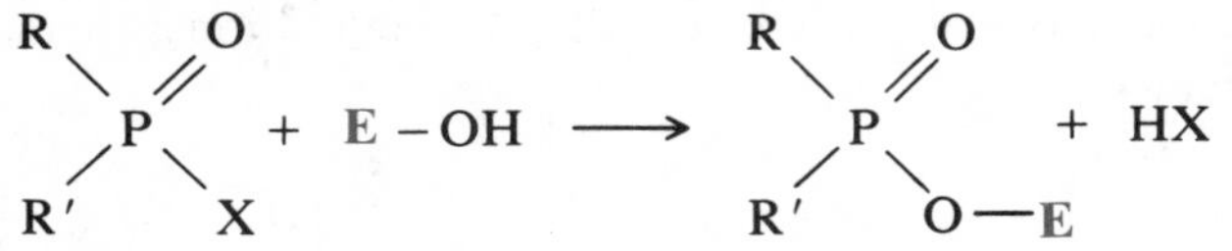

(ii) No hydrolysis of phosphorylated enzyme complex

Figure 6.7   The mechanism of action of acetylcholinesterase.

Types of inhibition

COMPETITIVE

In this type, the inhibitor binds to specific groups at the active site to form an enzyme–inhibitor complex. The substrate and the inhibitor both compete for the same site, hence the name (Fig. 6.8a). This competition for the same site occurs because of the similarity in chemical structures of the inhibitor (Fig. 6.9) and the natural substrate. In the case of a competitive inhibitor, the inhibition can be relieved by increasing the concentration of substrate.

NON-COMPETITIVE

The non-competitive inhibitor does not bind to the active site but reacts at another site on the enzyme distorting the molecule and lowering the activity (Fig. 6.8b). In

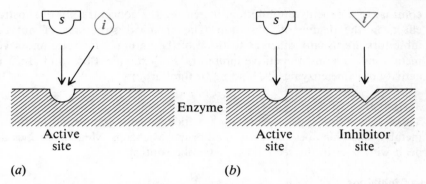

Active
site

Active
site

Inhibitor
site

Enzyme

(a)

(b)

Figure 6.8 Diagrammatic representation of the two main types of inhibition. (a) Competitive: substrate s and inhibitor i both compete for the active site. (b) Non-competitive: substrate s and inhibitor i bind to different sites.

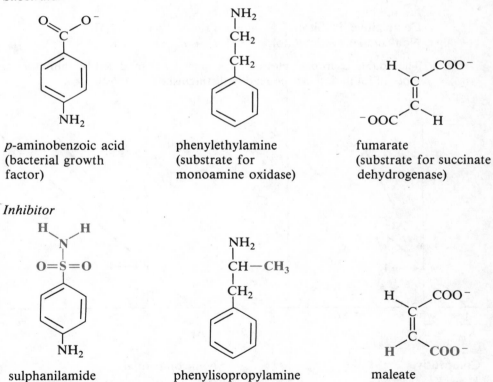

Substrate

p-aminobenzoic acid
(bacterial growth
factor)

phenylethylamine
(substrate for
monoamine oxidase)

fumarate
(substrate for succinate
dehydrogenase)

Inhibitor

sulphanilamide
(antibacterial agent)

phenylisopropylamine
(antidepressant drug)

maleate
(blocks the citric
acid cycle)

Figure 6.9 Some examples of competitive inhibitors.

contrast to competitive inhibition, increasing the concentration of substrate has no effect on the degree of inhibition. The chemical structures of non-competitive inhibitors are usually different to those of the natural substrates and several heavy metal ions are non-competitive inhibitors. For example, Pb^{2+} and Hg^{2+} inhibit the activity of some enzymes by binding to thiol groups.

ALLOSTERIC
There is a third type of inhibition which is very important in the control of metabolism. This does not obey the normal Michaelis–Menten kinetics and will be dealt with later in the section on metabolic control.

The kinetics of inhibition
The reaction velocity of an enzyme depends on the concentration of substrate as discussed earlier and in the presence of an inhibitor concentration i, the activity is changed according to the equations below where K_1 is the enzyme–inhibitor equilibrium constant.

No inhibitor: $v = V/(1 + K_m/s)$
Competitive inhibitor: $v = V/[1 + K_m/s(1 + i/K_1)]$
Non-competitive inhibitor: $v = V/(1 + K_m/s)(1 + i/K_1)$

The double reciprocal plot of $1/v$ against $1/s$ is modified by inhibitors and the two types of inhibition can be readily distinguished (Fig. 6.10).

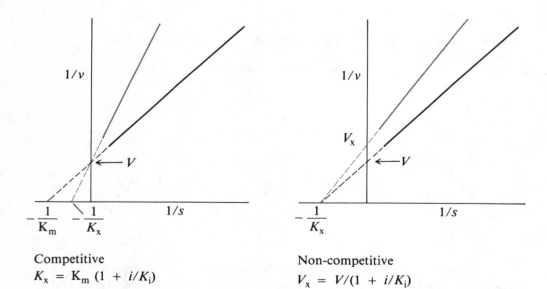

Competitive
$K_x = K_m (1 + i/K_i)$

Non-competitive
$V_x = V/(1 + i/K_i)$

Figure 6.10 Double reciprocal plots for competitive (black) and non-competitive (red) types of inhibition.

6.4 Variation of enzyme activity: pH, temperature and cofactors

Enzymes will only function under narrowly defined chemical and physical conditions and enzyme assays should always be carried out under conditions that maximize the activity of the enzyme. The importance of substrate and inhibitors has already been considered but there are factors apart from these that affect the activity of an enzyme.

pH

OPTIMUM PH

Enzymes are only active over a limited pH range and a plot of activity against pH gives a bell-shaped curve for most enzymes similar to that illustrated in black in Fig. 6.11.

The pH value where the activity is a maximum is known as the *optimum pH* and is due to the presence of charged groups on the enzyme protein. The ionization of these groups varies with pH and only one of the many possible ionized forms is enzymatically active.

Many enzymes have an optimum between pH 5 and pH 7 although some are active outside this range. The two extremes are probably pepsin, with an acid optimum of about pH 2, and alkaline phosphatase, which functions best under alkaline conditions at pH 10.

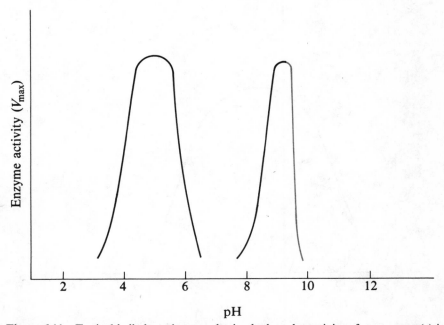

Figure 6.11 Typical bell-shaped curve obtained when the activity of an enzyme (v) is plotted against pH (black, stable enzyme; red, enzyme unstable at alkaline pH).

pH AND ENZYME STABILITY

All enzymes are adversely affected by extreme pH values which denature the enzyme protein but some enzymes begin to lose their activity at a relatively moderate pH. In these cases, the pH–activity curve is not symmetrical and the activity falls rapidly either side of the optimum due to denaturation (see red curve in Fig. 6.11).

pH AND V

If log V is plotted against pH, a curve is obtained which consists of three straight lines joined by two small curves (Fig. 6.12). The intercepts of these three lines give the pK_a values of the ionizable groups of the ES complex and these groups can be identified by comparing the pK_a values obtained with those for the amino acid residues present in peptides and small proteins.

pH AND K_m

If pK_m ($-\log_{10} K_m$) is plotted against pH a similar plot is obtained to Fig. 6.12 but with more breaks. In this case, the intercepts of the straight line parts of the curve give the pK_a values of the ionizable groups on the substrate as well as those of the ES complex.

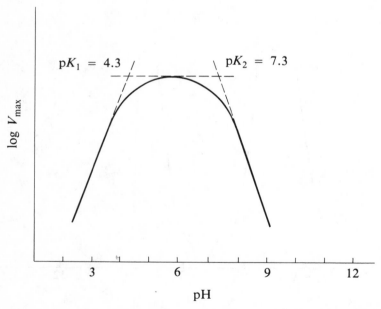

Figure 6.12   A plot of log V against pH to obtain the pK_a values of the ionizing groups of the ES complex.

Temperature

'OPTIMUM TEMPERATURE'

Unlike optimum pH, enzymes do not have an 'optimum temperature'. As the temperature rises there is an increase in the reaction velocity but there is also an increase in the rate at which the enzyme is destroyed. The so-called optimum temperature is a result of these two opposing effects and depends on the length of time over which the measurements are made: the shorter the time, the higher the apparent optimum temperature; the longer the time, the lower the apparent optimum temperature.

INACTIVATION OF ENZYMES

Nearly all enzymes are stable at 37 °C but high and sometimes low temperatures lead to their inactivation.

High temperature   Enzymes, like all proteins, are denatured by high temperatures and most enzymes lose their activity between 50 and 70 °C. However, a few are stable at high temperature but they are the exception to this general rule. Ribonuclease, for example, can be boiled under acid conditions and regains its enzyme activity when cooled, while the α-amylase of bacteria found in hot springs loses only 10 per cent of its activity in 1 h at 90 °C.

Low temperature   Nearly all enzymes are stable at low temperatures and, in the laboratory, tissues and cell fractions are kept at 0 °C to preserve the activity of the preparations. However, a small minority of enzymes are actually cold-labile due to the loss of cofactors or the dissociation of subunits. Mitochondrial ATPase is one example of an enzyme that is stable at room temperature but undergoes rapid inactivation at 0 °C.

TEMPERATURE AND ENZYME ACTIVITY

Towards the end of the last century, Arrhenius obtained the empirical equation for a chemical reaction:

$$\mathrm{d} \ln k / \mathrm{d}T = E/RT^2$$

where k = rate constant, T = temperature (K), R = gas constant and E = energy of activation.

He postulated that when a chemical reaction takes place an activated complex is first formed and that an energy barrier has to be overcome before the reaction can proceed. The energy required to give the activated complex is the *energy of activation* (Fig. 6.1).

Integrating this equation we obtain:

$$\ln k = -E/RT + \text{constant}$$

The energy of activation of any chemical reaction can therefore be obtained from a

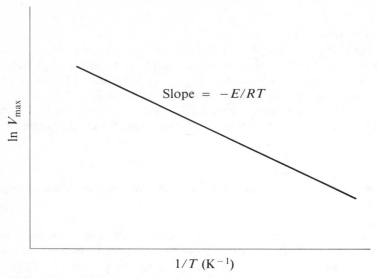

Figure 6.13 Arrhenius plot for an enzyme-catalysed reaction.

plot of $\ln k$ against $1/T$. In the case of enzyme-catalysed reactions, $V = k \times$ total enzyme concentration so $\ln V$ is plotted against the reciprocal of the absolute temperature. This is known as an *Arrhenius plot* and the energy of activation can be obtained from the slope of the straight line (Fig. 6.13). A more fundamental derivation has shown that:

$$E = \Delta H + RT$$

where ΔH is the *enthalpy* of the reaction.

Coenzymes, prosthetic groups and activators

HISTORY

During the early work on fermentation it was noticed that an extract of yeast was able to ferment glucose but not if it was dialysed. The fermentation could be recovered by adding back the dialysate which is the low-molecular-weight fraction. The ability to ferment could also be regained even if the dialysate was boiled, then cooled before mixing with the dialysed extract. It therefore became clear that fermentation needed not only enzymes, which are non-dialysable and thermolabile, but also low-molecular-weight compounds, which are thermostable. We now know that this low-molecular-weight fraction consists of a mixture of low-molecular-weight coenzymes and activators.

COENZYMES

These are soluble compounds which are needed for enzyme activity and which take an active part in the mechanism of catalysis, often acting as donors or acceptors of specific chemical groups.

NAD   Nicotinamide adenine dinucleotide (NAD) is

$$\text{nicotinamide—ribose—phosphate}$$
$$|$$
$$\text{adenine—ribose—phosphate}$$

a coenzyme that accepts and donates hydrogen atoms and is an essential cofactor for dehydrogenase enzymes such as lactate dehydrogenase (LDH).

$$CH_3CH(OH)COO^- + NAD^+ \overset{[LDH]}{\rightleftharpoons} CH_3CO \cdot COO^- + NADH + H^+$$

$$\text{lactate} \qquad\qquad\qquad \text{pyruvate}$$

CoA-SH   Another important coenzyme is coenzyme A (CoA-SH) or adenine-ribose-Ⓟ-Ⓟ-D-pantothenic acid-β-mercaptoethylamine. The sulphydryl group (—SH) at the end of the molecule carries the acyl groups as thioesters so coenzyme A is usually written CoA-SH and acetyl coenzyme A as CH_3COO-S-CoA.

PROSTHETIC GROUPS

This term is used for coenzymes that are tightly bound to the enzyme protein and are part of the enzyme molecule. Flavoproteins are prosthetic groups and FAD accepts and donates hydrogen atoms in redox reactions such as succinate dehydrogenase (Section 6.3). The distinction between coenzymes and prosthetic groups is not rigid and prosthetic groups can sometimes be removed from the enzyme protein given the right conditions.

Table 6.3 Some B-group vitamins as precursors of coenzymes and prosthetic groups

Vitamin (coenzyme or prosthetic group)	Metabolic role of cofactor
Thiamine (B_1) (thiamine pyrophosphate, TPP)	Cofactor of oxidative decarboxylations in the Krebs' cycle.
Riboflavin (B_2) (flavin adenine dinucleotide, FAD)	Prosthetic group of some dehydrogenases.
Pyridoxal (B_6) (pyridoxal phosphate)	Coenzyme for aminotransferase and decarboxylase reactions of amino acid metabolism.
Niacin (nicotinamide adenine dinucleotide, NAD)	Coenzyme for a large number of dehydrogenases.
Pantothenic acid (coenzyme A, CoA)	An essential cofactor for the oxidation of fats and carbohydrates in the Krebs' cycle.
Biotin (biotin)	Coenzyme for reactions involving fixation of carbon dioxide (carboxylases).

Vitamins and coenzymes

Vitamins are organic molecules that the organism is unable to synthesize in sufficient amounts for its needs and small quantities must be supplied in the diet to maintain normal health. The biochemical basis for the requirement of many vitamins is still not fully understood but many of the B group of vitamins are necessary for health because they are the precursors of coenzymes and prosthetic groups (Table 6.3).

ACTIVATORS

These have a very simple chemical structure and are generally far less specific than coenzymes. Several metal ions are known to be activators and Mg^{2+} is needed for many enzymes involving the addition (kinases) or removal (phosphatases) of phosphate groups while Mn^{2+} is an activator for a large number of enzymes and is often interchangeable with Mg^{2+}. Enzymes are not normally affected by anions and salivary amylase is unusual in that the activity is increased Cl^- and other anions.

Part II   Metabolism: the chemical changes in living matter

7. Energy and life

7.1 Principles of thermodynamics: to be or not to be

The structure and function of living matter is highly organized and such an ordered state can only be maintained by the supply of energy which arises from the oxidation of food. Living matter needs energy for many vital processes (Fig. 7.1) and the factors that affect the capture, utilization and interconversion of metabolic energy are the subjects of this section.

The study of the conversion of energy to its different forms and its interaction with matter is the science of *thermodynamics* and although the pioneer work was carried out in physics and engineering, the basic principles can also be applied to living systems. Thermodynamics is not an esoteric subject of academic interest only but describes whether or not a reaction can occur and this can be literally a matter of life or death for a cell. The nature of this book means that only the basic principles of the subject can be dealt with and no attempt has been made to derive the equations of thermodynamics. The reader is therefore recommended to read the relevant chapters

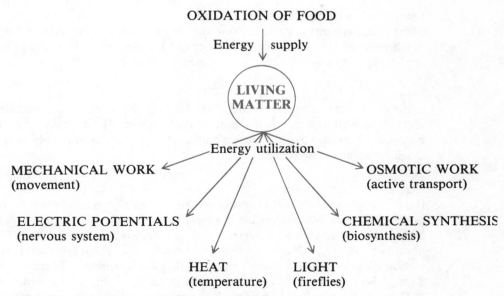

Figure 7.1 Energy turnover in living organisms.

in one of the many excellent books on physical chemistry for the biological sciences to see how these equations arise from first principles.

The first law of thermodynamics

The first law is the law of the conservation of energy and states that: '*In a chemical reaction, energy cannot be created or destroyed.*' The form of the energy can be changed but not its total quantity so that another way of describing this law is to say that: '*The total energy of a system and its surroundings is constant.*'

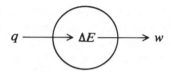

If heat flows into a system (q) and work is done on the environment (w) then the internal energy of the system is increased by ΔE so that:

$$\Delta E = q - w$$

The symbol Δ placed before a quantity describes the increase or decrease in that quantity following the change in the state of the system. In the case of a chemical reaction:

$$\Delta E = \Delta E \text{ (products)} - \Delta E \text{ (reactants)}$$

DEFINITION OF ENTHALPY (ΔH)

Most living systems operate at constant temperature and pressure and under these conditions, the change in internal energy (ΔE) is the same as the change in the total heat energy or *enthalpy* (ΔH), so that $\Delta E = \Delta H$ and

$$\Delta H = q - w \qquad (1)$$

If no work is done by the system then $\Delta H = q$, the whole of the enthalpy change appears as heat and ΔH is the heat of the reaction. The enthalpy therefore is the maximum heat energy of a system but as we shall see below, only part of this is available to do useful work.

ENTHALPY OF A REACTION

If ΔH is negative then heat is produced and the reaction is *exothermic* while if ΔH is positive then heat is taken up and the reaction is *endothermic*. If glucose is oxidized to carbon dioxide and water, a large amount of energy is released so that ΔH^0 is a large negative value (Fig. 7.2) and the reaction is exothermic. The reverse reaction, in which carbon dioxide and water are converted to glucose with the liberation of oxygen during photosynthesis, has a high positive enthalpy change and is endothermic.

Thermodynamics is not concerned with the detailed mechanisms that take place in

$$C_6H_{12}O_6 + 6O_2 \longrightarrow 6CO_2 + 6H_2O$$

Molecules   :   1 + 6 $\longrightarrow$ 6 + 6

State   :   solid + gas $\longrightarrow$ gas + liquid

ΔH_f° (kJ/mol) : -912 0 $\longrightarrow$ -393 -238

ΔH° for the above reaction is therefore:

$$\Delta H^\circ = [\Delta H^\circ \text{ products}] - [\Delta H^\circ \text{ reactants}]$$

$$\Delta H^\circ = [-6 \times (-393) + 6 \times (-238)] - [1 \times (-912) + 0]$$

$$\Delta H^\circ = -2874 \text{ kJ/mol}$$

Figure 7.2   Changes in entropy and enthalpy during the oxidation of glucose.

a system but only with the overall change from one state to another. This means that the energy released during the burning of glucose in a bomb calorimeter is the same as the energy released during the oxidation of glucose in the cell. ΔH^0 therefore is a constant and is independent of the pathway taken.

Second law of thermodynamics

Many processes will occur quite naturally and are described as spontaneous. The use of the word spontaneous in thermodynamics can cause some confusion, since it does not mean that the process happens 'at once' but it indicates that a reaction will proceed in a particular direction when left to itself. For example, sugar crystals dissolve in a cup of tea, heat flows along a metal bar with one end in a flame and an ice cube melts in the hand. These are all spontaneous changes which take place in one direction only and they do not happen in reverse. Sugar crystals do not appear from sugar dissolved in hot tea, one end of an iron bar does not become hot by itself and an ice cube does not form from water held in the hand. The reason why these processes take place in one direction only is governed by the second law of thermodynamics which states that: '*Systems tend to proceed from an ordered to a more disordered state.*' Another way of describing this is to say that: '*The degree of disorder of the universe or a closed system is always increasing.*'

ENTROPY

The driving force for spontaneous processes is *entropy* which is a measure of the degree of randomness of the system. The change in entropy (ΔS) can be positive or negative and reactions with a positive ΔS tend to proceed spontaneously while those with a negative ΔS cannot take place without an input of energy. An example of the increase in entropy during a reaction can be seen during the oxidation of glucose, in which there is a change in state from a solid and a gas to one of a liquid and a gas and an increase in the number of molecules from 7 to 12 (Fig. 7.2). Mathematically, the change in entropy ΔS is the heat input q divided by the absolute temperature T so that:

$$\Delta S = q/T \tag{2}$$

Free energy

DEFINITION

If the first (1) and second (2) laws are combined then:

$$\Delta H = T \cdot \Delta S - w \quad \text{or} \quad -w = \Delta H - T \cdot \Delta S$$

At a constant pressure and temperature, the term w can be replaced by the change in the *free energy* ($\Delta G = -w$) so that for biochemical reactions:

$$\Delta G = \Delta H - T \cdot \Delta S$$

The *change in the free energy* (ΔG) is the maximum amount of energy that is available to do useful work and this is always less than the total heat energy or *enthalpy* (ΔH) of a system at all temperatures above absolute zero because of the *entropy* change (ΔS).

STANDARD FREE ENERGY CHANGE

This is the energy change of a reaction measured under standard conditions which are defined as 1 mol of a compound as an ideal gas or pure solid or liquid at a pressure of 1 atmosphere and a temperature of 25 °C (298 K). In the case of a solution, the standard state is a concentration of 1 mol/litre.

ΔG_f^0 This is the standard free energy of formation and is the energy needed to produce 1 mol of a compound from its elements in their standard state when the ΔG_f^0 for the elements is taken as zero.

ΔG^0 This is the standard free energy change of a reaction which can be calculated from the individual values of the reactants so that:

$$\Delta G^0 = \Delta G^0 \text{ (products)} - \Delta G^0 \text{ (reactants)}$$

$\Delta G^{0\prime}$ Many biochemical reactions involve H^+ and the standard state means that H^+ are present at 1 mol/litre with a pH of 0. This is far removed from the actual situation in the cell and so for biochemical reactions the standard state for H^+ is taken as pH 7 and the symbol $\Delta G^{0\prime}$ used instead of ΔG^0.

7.2 Bioenergetics: thermodynamics applied to biology

Bioenergetics is the study of energy transformation in living matter using the basic principles of thermodynamics. However, there are some fundamental differences between non-living and living systems which should always be borne in mind.

Free energy changes in living matter

FREE ENERGY AND CONCENTRATION

Firstly, reactions in living matter are at a steady state and not at equilibrium and concentrations are much less than molar. This means that the free energy changes of reactions *in vivo* can be quite different to the $\Delta G^{0\prime}$ values calculated *in vitro* using molar concentrations.

For a chemical reaction when

$$A + B \rightleftharpoons C + D$$

the free energy change is given by:

$$\Delta G = \Delta G^{0\prime} + RT \ln [C][D]/[A][B]$$

where R = gas constant and T = absolute temperature.

At equilibrium, ΔG is zero and the *mass action ratio* $[C][D]/[A][B] = K$, the equilibrium constant, so that:

$$\Delta G^{0\prime} = -RT \ln K$$

However, if the cell reactions are not at equilibrium, the relative values of the mass action ratio (MAR) and the equilibrium constant will determine the direction that a reaction takes.

Under standard conditions:

MAR and K	Direction of reaction	$\Delta G^{0\prime}$	Energy change
MAR = K	Equilibrium	0	zero
MAR < K	⟵――――――	+	endergonic
MAR > K	――――――⟶	−	exergonic

FREE ENERGY AND OTHER FACTORS

The thermodynamic equations used in biochemistry assume that the pressure and temperature remain constant and this may not always be true. Furthermore, changes in pH or the concentration of metal ions in the cell may cause a shift in the rate of reaction or the equilibrium position of reactions that are measured in the laboratory.

ENERGY OF ACTIVATION

The complete oxidation of glucose to carbon dioxide and water has a very high negative ΔG but, in spite of this, the oxidation is not spontaneous. Glucose and many other biomolecules are actually quite stable because of the energy of activation which has to be overcome before any reaction can take place (Fig. 7.3). In the laboratory the

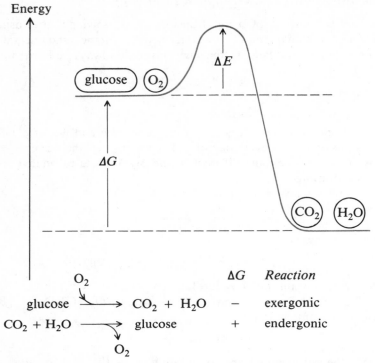

Figure 7.3 The energy of activation and free energy changes for the oxidation and synthesis of glucose.

barrier can be crossed by raising the temperature which increases the kinetic energy of the reactant molecules. However, such a procedure is not possible in living organisms which function at low ambient temperatures or at a constant 37 °C in the case of warm-blooded animals. In living matter, the energy barrier is overcome by lowering it using enzymes and in some cases by raising the energy of a reactant molecule by phosphorylation.

Adenosine triphosphate (ATP)

Energy must be supplied to sustain life and a human being with a sedentary occupation needs about 8000 kJ per day, while someone who carries out heavy manual work requires up to 16 000 kJ per day.

The energy demands throughout the day can be quite variable and depend very much on the activity of the organism. The resting or basal metabolic rate in humans is about 300 kJ/h for females and up to 360 kJ/h for males, which is the power output of a 100-watt light bulb. The metabolic rate doubles during light activity, increases some four times in moderate exercise such as swimming and can rise to more than ten times the basal rate when running a marathon. The energy for all this activity comes from the oxidation of food but more immediately from the breakdown of adenosine triphosphate (Fig. 7.4) which can be regarded as the *energy currency* of the cell.

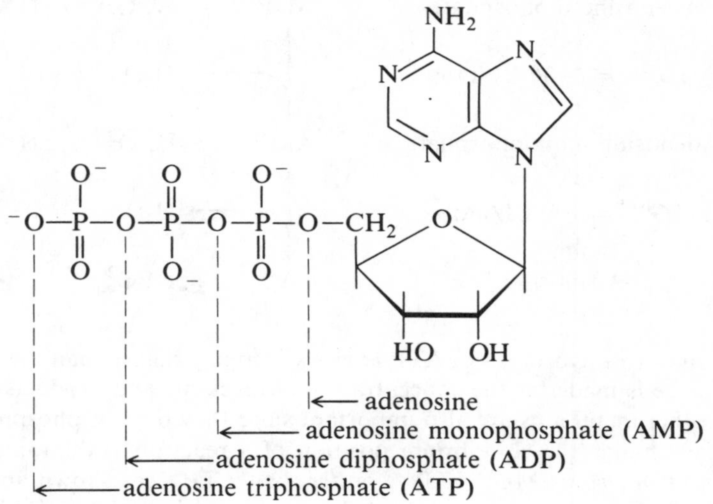

Figure 7.4 The structure of adenosine triphosphate (ATP) and related compounds.

ATP is 'earned' from energy-producing reactions such as photosynthesis or oxidations and 'spent' in a whole range of energy-utilizing reactions (Fig. 7.1). ATP is not a 'store' of energy but rather the link between exergonic and endergonic reactions.

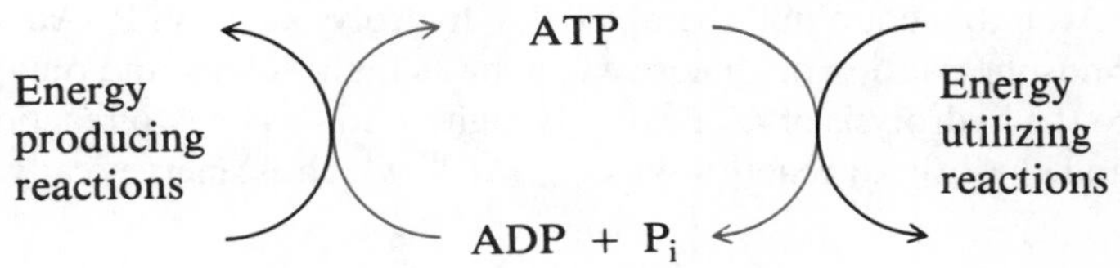

The turnover of this molecule in the cell is therefore very high and even in a resting human this amounts to about 0.5 g per second.

THE HYDROLYSIS OF ATP

All the phosphate groups of ATP can be hydrolysed with the liberation of energy. The standard free energy change for the hydrolysis of the first two phosphate groups is quite high and this gave rise to the totally mistaken idea of *high energy phosphate bonds*. This particular expression is still found in some texts but should not be used since the energy does not reside in any particular bond.

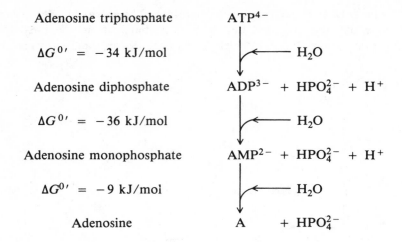

Adenosine triphosphate ATP^{4-}

$\Delta G^{0\prime} = -34$ kJ/mol

Adenosine diphosphate $ADP^{3-} + HPO_4^{2-} + H^+$

$\Delta G^{0\prime} = -36$ kJ/mol

Adenosine monophosphate $AMP^{2-} + HPO_4^{2-} + H^+$

$\Delta G^{0\prime} = -9$ kJ/mol

Adenosine $A + HPO_4^{2-}$

The free energy of hydrolysis of ATP *in vivo* is actually higher than the $\Delta G^{0\prime}$ value when allowance is made for the concentration of reactants and products in the cell. Mg^{2+} and other metal ions are also important since they bind to phosphate groups and this may change the equilibrium position of a reaction. ATP, for example, is largely present *in vivo* as $Mg^{2+}ATP^{4-}$ rather than ATP^{4-} as shown above. When allowance is made for these factors, the ΔG value for the hydrolysis of ATP to ADP under cellular conditions is about 50 kJ/mol.

ATP COUPLED REACTIONS

Many reactions are not spontaneous and cannot take place without an input of energy. Since many living organisms function at a constant temperature, energy cannot be supplied in the form of heat and energetically unfavourable reactions are driven by coupling them to the hydrolysis of ATP. An example of this is the phosphorylation of glucose which has a high $+\Delta G^{0\prime}$ and only takes place when linked to the hydrolysis of ATP with its high $-\Delta G^{0\prime}$. The two reactions are coupled together and the overall reaction has a $-\Delta G^{0\prime}$ which is energetically favourable.

	$\Delta G^{0\prime}$ (kJ/mol)
glucose + P_1 $\rightarrow$ glucose-6-phosphate + H_2O	+14
ATP + H_2O $\rightarrow$ ADP + P_1	−31
glucose + ATP $\rightarrow$ glucose-6-phosphate + ADP	−17

ATP therefore is the link between the energy-yielding reactions (catabolism) and the energy-requiring reactions (anabolism) of cellular metabolism and how ATP is generated and used is the subject of the next few sections.

7.3   Photosynthesis: light and life

The nature of photosynthesis

THE CAPTURE OF SOLAR ENERGY

The primary source of energy in the biosphere is the sun and all living organisms depend on the ability of plants and certain micro-organisms to capture and store the visible radiant energy of sunlight. During this process, known as *photosynthesis*, the energy of light is used to synthesize carbohydrates from the simple molecules of carbon dioxide and water. The overall reaction is commonly written as:

$$6CO_2 + 6H_2O \longrightarrow C_6H_{12}O_6 + 6O_2$$

although the final product of photosynthesis is not a hexose sugar as shown above but usually sucrose or starch.

In effect, solar energy is captured and stored as chemical energy in the organic molecules synthesized. This energy is then released when the carbohydrates are oxidized in the plants or by animals after being eaten and digested (Fig. 2.7).

CARBON FIXATION

During photosynthesis, carbon is transferred from the atmosphere to the biosphere and about 10^{10} tonnes of the element are fixed in this way each year. About 90 per cent of photosynthesis takes place in the micro-organisms in the oceans of the world and only 10 per cent on land. However, plants are an important part of the human environment and a valuable source of food and raw materials so that interest has tended to concentrate on photosynthesis in green plants.

The absorption of light

The first step in photosynthesis is the absorption of light by coloured pigments and the effectiveness of different wavelengths of light at promoting photosynthesis closely follows the absorption spectrum of the pigments extracted from leaves.

CHLOROPHYLLS

The most important of these pigments are the chlorophylls and there are a number of these compounds found in nature. Plants contain two of these pigments, one of which is always chlorophyll a and the structure of this compound is shown in Fig. 7.5. The conjugated double bond system of the porphyrin ring is responsible for the high absorption of light from the visible part of the spectrum and chlorophylls absorb strongly at the blue and the red ends of the visible spectrum (Fig. 7.6). Chlorophyll therefore transmits light in the middle of the spectral range which is why plants appear green.

CAROTENOIDS

Plants also contain carotenoids which are red and yellow compounds and which therefore absorb light from a different part of the spectrum to that of the chlorophylls.

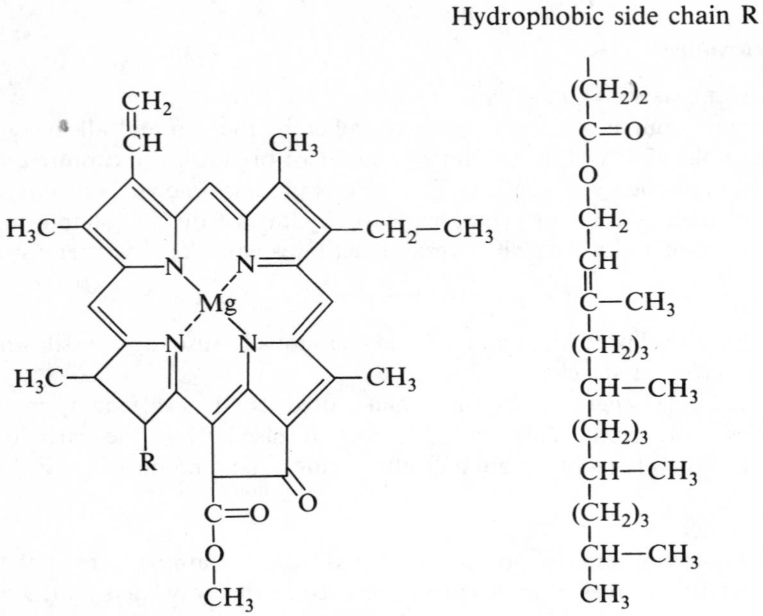

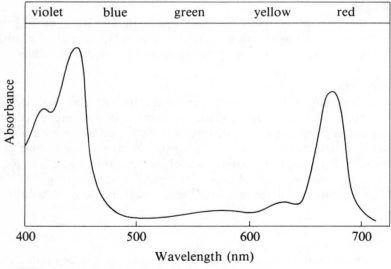

Figure 7.5   The structure of chlorophyll a.

Figure 7.6   The absorption spectrum of chlorophyll a.

The main function of these compounds is to absorb light from a region of the spectrum where chlorophyll absorbs only poorly and to transfer the energy to chlorophyll. Carotenoids also serve to protect the cell against damage from oxygen free radicals when high concentrations of oxygen are produced at high light intensities.

The production of oxygen

THE HILL REACTION

Light is absorbed by a large number of chlorophyll molecules and this energy is transferred to a few special molecules of chlorophyll located at the reaction centre and it is here that the radiant energy is converted into chemical energy. The energy absorbed at the reaction centre removes electrons from the chlorophyll and leaves a cationic free radical which is a powerful reductant. The electrons removed are then replaced by electrons from water and this *photolysis of water* liberates molecular oxygen and protons. This evolution of oxygen and the simultaneous reduction of an electron acceptor was discovered by Hill using an artificial acceptor and the same type of reaction takes place in the cell with $NADP^+$ as the electron acceptor *in vivo*.

Electron acceptor

(a) Artificial $\quad 2H_2O + 4Fe^{3+} \longrightarrow O_2 + 4H^+ + 4Fe^{2+}$

(b) Natural $\quad 2H_2O + 2NADP^+ \longrightarrow O_2 + 2H^+ + 2NADPH$

The light effectively causes the electrons to flow 'uphill' and this is then followed by a flow 'downhill' in which the electrons pass through a series of compounds in the form of an electron transport chain in which the components are alternatively reduced and oxidized. During the flow of electrons through the system, $NADP^+$ is reduced to NADPH and H^+ and a proton gradient is formed across the thylakoid membrane of the chloroplast. The discharge of this proton gradient then gives rise to ATP in a similar way to that found in mitochondria (Section 7.6).

THE Z SCHEME

In plants there are actually two light-driven reactions which take place at two centres, photosystem I (PSI) and photosystem II (PSII). These two systems operate together in the form of a zigzag or Z scheme and the essential features of this electron transport chain are shown in Fig. 7.7.

Light is absolutely essential for the production of oxygen and the generation of reducing power so the first part of photosynthesis is known as the *light reaction*. The next part of photosynthesis, in which the NADPH and ATP are used, does not need light and the details of the *dark reaction* are discussed in Section 7.4.

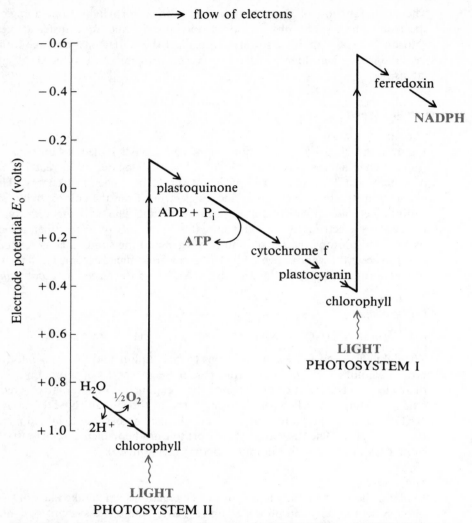

Figure 7.7 Electron transport and the generation of NADPH and ATP in chloroplasts.

7.4 Fixation of carbon: the dark reaction of photosynthesis

The dark reaction

AN OVERVIEW

ATP and reducing equivalents in the form of MADPH are generated as electrons flow through the electron transport system in the *light reaction* (Fig. 7.7). These compounds are then used to reduce carbon dioxide and form carbohydrate in the next stage of photosynthesis which is known as the *dark reaction*.

EARLY EXPERIMENTS

The classical experiments that led to the details of the dark reaction were carried out in Calvin's laboratory and are an excellent example of how radioactive isotopes can be used to trace a metabolic pathway.

During photosynthesis, the green algae *Chlorella* were exposed to carbon dioxide labelled with radioactive carbon (^{14}C) and at the end of the experiment the cells were killed and the radioactive products separated and identified by paper chromatography. Different periods of illumination were tried and this enabled the order in which the various products were formed to be determined. Further experiments were made by other researchers who identified the enzymes involved in the metabolic pathway which is now called the *Calvin cycle*.

The Calvin cycle

THE INCORPORATION OF CARBON DIOXIDE INTO 3-PHOSPHOGLYCERATE

If the *Chlorella* were illuminated for just a few seconds then only one compound was found to be labelled, namely 3-phosphoglycerate and this is now recognized to be the first product of photosynthesis. The carbon dioxide combines with the five-carbon compound ribulose-1,5-bisphosphate to form a transient six-carbon intermediate which is then hydrolysed to two molecules of 3-phosphoglycerate:

$$CO_2 + H_2O + \begin{array}{c} CH_2O\,\text{\textcircled{P}} \\ | \\ C=O \\ | \\ H-C-OH \\ | \\ H-C-OH \\ | \\ CH_2O\,\text{\textcircled{P}} \end{array} \xrightarrow{[RBC]} \begin{array}{c} CH_2O\,\text{\textcircled{P}} \\ | \\ H-C-OH \\ | \\ COO^- \end{array} + \begin{array}{c} COO^- \\ | \\ H-C-OH \\ | \\ CH_2O\,\text{\textcircled{P}} \end{array}$$

	ribulose-1,5-bisphosphate	3-phosphoglycerate
1C	5C	2 × 3C

This reaction is catalysed by the enzyme ribulose bisphosphate carboxylase (RBC) which is probably the most abundant enzyme in the world.

THE FORMATION OF GLYCERALDEHYDE-3-PHOSPHATE

The next stage in the Calvin cycle uses ATP and NADPH generated in the light reaction to synthesize 3-phosphoglyceraldehyde from the 3-phosphoglycerate. The first step is the phosphorylation of the 3-phosphoglyceraldehyde by ATP to form 1,3-bisphosphoglycerate, a reaction catalysed by the enzyme phosphoglycerate kinase (PGK):

$$
\begin{array}{ccc}
\text{CH}_2\text{O}\,\textcircled{P} & & \text{CH}_2\text{O}\,\textcircled{P} \\
| & & | \\
\text{H} - \text{C} - \text{OH} + \text{ATP} \xrightarrow{\text{[PGK]}} & \text{H} - \text{C} - \text{OH} + \text{ADP} \\
| & & | \\
\text{COO}^- & & \text{COO}\,\textcircled{P}
\end{array}
$$

3-phosphoglycerate   1,3-bisphosphoglycerate

The second stage is the reduction of this compound by NADPH to form 3-phosphoglyceraldehyde and this is catalysed by the enzyme glyceraldehyde-3-phosphate dehydrogenase (GPDH):

$$
\begin{array}{ccc}
\text{CH}_2\text{O}\,\textcircled{P} & & \text{CH}_2\text{O}\,\textcircled{P} \\
| & & | \\
\text{H} - \text{C} - \text{OH} + \text{NADPH} + \text{H}^+ \xrightarrow{\text{[GPDH]}} & \text{H} - \text{C} - \text{OH} + \text{P}_i + \text{NADP}^+ \\
| & & | \\
\text{COO}\,\textcircled{P} & & \text{CHO}
\end{array}
$$

1,3-bisphosphoglycerate   3-phosphoglyceraldehyde

REACTIONS OF THE PENTOSE PHOSPHATE AND GLYCOLYTIC PATHWAYS

After this the reactions become complicated and a number of interconversions take place with 3-phosphoglyceraldehyde as the starting point. Three molecules of carbon dioxide and three molecules of ribulose-1,5-bisphosphate produce six molecules of 3-phosphoglyceraldehyde. Five of these molecules undergo a series of transformations to regenerate ribulose-1,5-bisphosphate while the sixth, 3-phosphoglyceraldehyde, is the starting point for the synthesis of sucrose, starch and other plant carbohydrates (Fig. 7.8).

THE NET RESULT OF THE CYCLE

A simplified version of the Calvin cycle is given in Fig. 7.9 and shows how ATP and NADPH are used to reduce carbon dioxide and produce 3-phosphoglyceraldehyde, the precursor of the plant carbohydrates. The cycle can therefore be summarized as follows:

$$3\text{CO}_2 + 9\text{ATP} + 6\text{NADPH} + 6\text{H}^+ + 5\text{H}_2\text{O}$$

$$\downarrow$$

3-phosphoglyceraldehyde $+ 9\text{ADP} + 8\text{P}_1 + 6\text{NADP}^+$

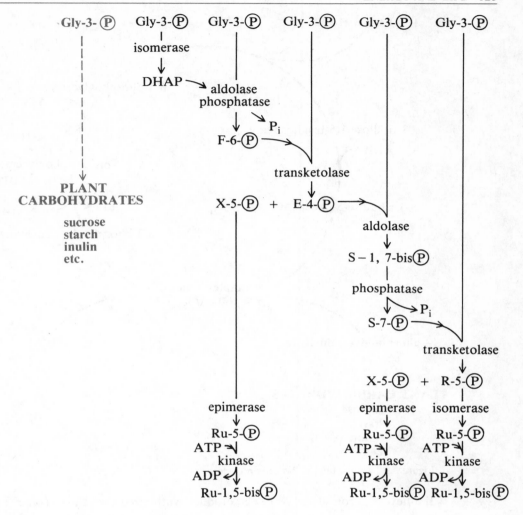

Figure 7.8   Reactions of the Calvin cycle.

	Intermediates	*No. of C atoms*
DHAP	= dihydroxyacetone phosphate	3
F-6-Ⓟ	= fructose-6-phosphate	6
E-4-Ⓟ	= erythrose-4-phosphate	4
X-5-Ⓟ	= xylulose-5-phosphate	5
S-1,7-bis Ⓟ	= sedoheptulose-1,7-bisphosphate	7
S-7-Ⓟ	= sedoheptulose-7-phosphate	7
R-5-Ⓟ	= ribose-5-phosphate	5
Ru-5-Ⓟ	= ribulose-5-phosphate	5
Ru-1,5-bis Ⓟ	= ribulose-1,5-bisphosphate	5

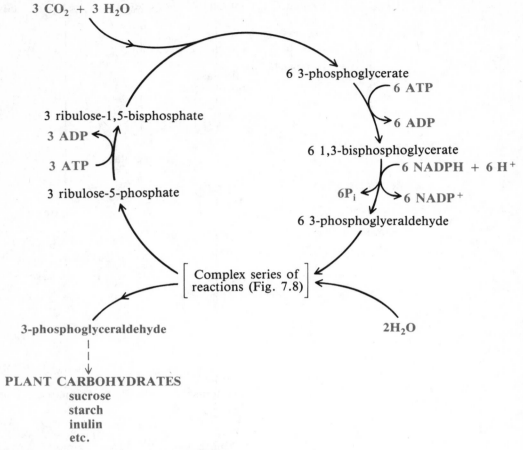

Figure 7.9   An outline of the Calvin cycle.

If a hexose is considered to be the product synthesized then this becomes:

$$6CO_2 + 18ATP + 12NADPH + 12H^+ + 12H_2O$$

$$\downarrow$$

$$C_6H_{12}O_6 + 18ADP + 18P_i + 12NADP^+$$

This is not exactly double the first equation since the extra two molecules of water and two of inorganic phosphate come from the action of fructose-1,6-bisphosphatase and glucose-6-phosphatase when 3-phosphoglycerate is converted to the hexose sugar.

In the plant cell, there is no net breakdown of ATP as the molecule is rapidly synthesized from ADP and P_i in the light reaction or by oxidation of metabolites. This reversal of esterification produces a molecule of water for every molecule of ATP formed so the net change in the dark reaction can be written:

$$12H_2O + 6CO_2 + 12NADPH + 12H^+ \longrightarrow C_6H_{12}O_6 + 12NADP^+ + 18H_2O$$

On cancelling this becomes:

$$6CO_2 + 12NADPH + 12H^+ \longrightarrow C_6H_{12}O_6 + 12NADP^+ + 6H_2O$$

A SUMMARY OF PHOTOSYNTHESIS

Reducing equivalents produced in the *light reaction* as NADPH are used to reduce carbon dioxide and form plant carbohydrates in the *dark reaction*. ATP is also formed by the light reaction and the breakdown of this compound to ADP and Pi provides the energy needed for the fixation of carbon dioxide.

Light reaction
$$12H_2O + 12NADP^+ \longrightarrow 6O_2 + 12NADPH + 12H^+$$

Dark reaction
$$6CO_2 + 12NADPH + 12H^+ \longrightarrow C_6H_{12}O_6 + 12NADP^+ + 6H_2O$$

Overall reaction
$$6CO_2 + 6H_2O \longrightarrow C_6H_{12}O_6 + 6O_2$$

7.5 Oxidation: the release of energy

The previous two sections showed that in the course of photosynthesis, energy is captured and carbon dioxide is reduced to carbohydrate. During respiration, this process is reversed and energy is liberated when carbohydrate is oxidized to carbon dioxide. Energy transfer in the living cell is associated with oxidation and reduction so it is important to understand the nature of these chemical processes.

Oxidation and reduction

REDOX REACTIONS

Oxidation and reduction involve the transfer of electrons. In *oxidation* there is a loss of electrons while during *reduction* there is a gain of electrons. There are similarities here to acids and bases which gain or lose a proton. An acid is always accompanied by its conjugate base and a reductant is always associated with an oxidant.

$$CH_3COOH \rightleftharpoons CH_3COO^- + H^+$$
$$\text{acid} \qquad\qquad \text{base} \qquad \text{proton}$$
$$Fe^{2+} \rightleftharpoons Fe^{3+} + e^-$$
$$\text{reductant} \qquad \text{oxidant} \quad \text{electron}$$

It is not possible to have a reductant without a corresponding oxidant so it is more accurate to talk about a *redox reaction*. In practice the electrons lost by one reductant during oxidation are taken up by another oxidant during reduction to give a *redox pair*.

Oxidation $\qquad\qquad Fe^{2+} \longrightarrow Fe^{3+} + e^-$

Reduction $\qquad\qquad e^- + Cu^{2+} \longrightarrow Cu^+$

Redox pair $\qquad\qquad Fe^{2+} + Cu^{2+} \longrightarrow Fe^{3+} + Cu^+$

TYPES OF OXIDATION

All oxidations involve the loss of electrons and this can take place in several ways.

Removal of electrons With some oxidations, electrons are transferred directly and an example of this is the oxidation of cytochromes when the oxidation state of the haem iron changes from ferrous to ferric.

$$\text{cytochrome } (Fe^{2+}) \longrightarrow \text{cytochrome } (Fe^{3+}) + e^- \text{ (to acceptor)}$$

Removal of hydrogen In most cases the loss of electrons is indirect and occurs when hydrogen is removed from a molecule. Electrons are lost during dehydrogenation, either as two electrons and two protons ($2e^- + 2H^+$) or as a hydride ion and a proton ($H^- + H^+$).

$$CH_3CH_2OH \longrightarrow CH_3CHO + 2H \text{ (to acceptor)}$$
$$\text{ethanol} \qquad\qquad \text{acetaldehyde}$$

Addition of oxygen   The addition of oxygen to a molecule is the most obvious form of oxidation but it may not be clear exactly how this involves the loss of electrons. Oxygen is a very electronegative element and attracts electrons very strongly so that the introduction of oxygen into a molecule leads to a redistribution of electrons in the structure as they are drawn towards this highly electronegative element and away from the rest of the atoms. Covalent bonds involving oxygen frequently show a small distribution of charge and it is this property which forms the basis of the hydrogen bond which is responsible for the cohesiveness of water (Section 2.2).

$$CH_3CHO + H_2O \longrightarrow CH_3COO^- + H^+ + 2H \text{ (to acceptor)}$$
$$\text{acetaldehyde} \qquad\qquad \text{acetate}$$

Biological oxidations

All oxidations fall within one of the three categories described above but those that take place in the cell are classified according to the type of enzyme involved and the mechanism of the oxidation.

OXYGENASES

These enzymes catalyse the direct addition of molecular oxygen to a substrate. The *di-oxygenases* catalyse the incorporation of both atoms of molecular oxygen into a structure while the *mono-oxygenases* introduce only one atom of oxygen into a substrate. The mono-oxygenases are also known as *hydroxylases* and require a cosubstrate to reduce the other oxygen atom to water. The oxidation of natural compounds by oxygenases is rare and most biological oxidations take place by other mechanisms. There are however some important reactions that are catalysed by oxygenases and one such example is the cleavage of the aromatic ring of catechol by catechol-1,2-dioxygenase (CDO). Degradative bacteria also contain dioxygenases and these open up the aromatic rings of the complex polymer *lignin*, an important component of wood.

$$\tfrac{1}{2}O_2 + \text{[benzene ring]} \xrightarrow{\text{[CDO]}} \text{[ring with COO}^-\text{, COO}^-\text{]} + 2H^+$$

OXIDASES

In these reactions, oxygen is not incorporated into the molecule but used as the acceptor of electrons to form hydrogen peroxide or water. A good example of this type of enzyme is cytochrome oxidase (CO) which is found in mitochondria and catalyses the oxidation of cytochrome c and protons to water by molecular oxygen. Most oxidases form hydrogen peroxide rather than water and cytochrome oxidase is unusual in that it is one of the few enzymes capable of reducing molecular oxygen to water:

$$2 \text{ cytochrome c } (Fe^{2+}) + O_2 + 4H^+$$
$$\Big\downarrow \text{[CO]}$$
$$2 \text{ cytochrome c } (Fe^{3+}) + 2H_2O$$

DEHYDROGENASES

Most biological oxidations take place by the removal of hydrogen rather than by the addition of oxygen and there are numerous examples of this in metabolism. In nearly all cases the oxidation is a two-electron transfer and a pair of hydrogens are removed and passed on to an acceptor. Alcohol features prominently in the life of mankind and the ability to break down ethanol determines its level in the blood and therefore its effect on our behaviour. Liver alcohol dehydrogenase (ADH) is an important enzyme and is a good example of a dehydrogenase that uses the coenzyme NAD^+ as the hydrogen acceptor.

$$CH_3CH_2OH + NAD^+ \xrightleftharpoons{[ADH]} CH_3CHO + NADH + H^+$$

Hydrogen and electron acceptors in living cells

THE NICOTINAMIDE NUCLEOTIDES (NAD AND NADP)

These compounds, which are coenzymes for many dehydrogenases, are dinucleotides and have the general structure:

$$\text{nicotinamide} - \text{ribose} - \textcircled{P}$$
$$|$$
$$\text{adenine} - \text{ribose} - \textcircled{P} \qquad \text{NAD and NADP}$$
$$|$$
$$\textcircled{P}$$

The structures of NAD nad NADP are the same except that NADP has an extra phosphate group on the 2'-position of the adenine ribose. During reduction, the only part of NAD or NADP that changes is the nicotinamide moiety which is derived from the B vitamin nicotinic acid. If R is the rest of the molecule and SH_2 is a substrate such as ethanol, the reduction can be written:

$$SH_2 + NAD^+ \xrightleftharpoons{} S + NADH + H^+$$

During reduction, one hydrogen is transferred as a hydride ion to the 4-position of the nicotinamide ring and the other hydrogen appears as a proton.

The ultraviolet absorption spectra of the oxidized and reduced forms of the coenzymes are different and the quinoid structure of NADH gives rise to a strong

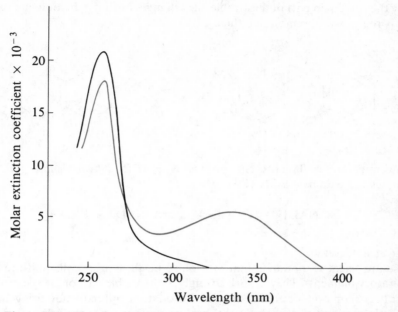

Figure 7.10   The ultraviolet absorption spectra of nicotinamide adenine dinucleotide (black, oxidized form NAD+; red, reduced form NADH).

peak at 340 nm which is not present in NAD^+ (Fig. 7.10). This property is used in the laboratory to assay dehydrogenases that use NAD or NADP as coenzymes and the rate of appearance or disappearance of NADH can be continuously followed by the change in absorption at 340 nm. The molar extinction coefficient of NADH at 340 nm is very high (6.3×10^3 litres mol^{-1} cm^{-1}) and enzyme activities involving changes of substrate as low as nanomoles per minute can be readily detected.

FLAVOPROTEINS (FMN AND FAD)

Some dehydrogenase enzymes have flavoprotein prosthetic groups as hydrogen acceptors which are flavins firmly bound to protein. They are known as flavin mononucleotide (FMN) and flavin adenine dinucleotide (FAD) but this is not strictly accurate since the pentose moiety of flavin is present as the alcohol ribitol rather than the sugar ribose. The flavins are derived from the vitamin riboflavin and have a yellow colour unlike the nicotinamide nucleotides which do not absorb in the visible region of the spectrum.

FMN   **FAD**

flavin — ribitol — Ⓟ   flavin — ribitol — Ⓟ

                              |

                    adenine — ribose — Ⓟ

Only the riboflavin part of these molecules changes during reduction and oxidation so if R is the rest of the molecule then:

$$\mathrm{_3HC}\!\!-\!\!\diagdown\!\!\diagup\!\!N\!\!-\!\!N\!\!=\!\!O \quad + 2H \rightleftharpoons \quad \mathrm{_3HC}\!\!-\!\!\diagdown\!\!\diagup\!\!N\!\!-\!\!N\!\!=\!\!O$$

An example of a flavoprotein enzyme is *NADH dehydrogenase* (NDH) which catalyses the oxidation of NADH to NAD^+.

$$NADH + H^+ + FAD \xrightleftharpoons{[NDH]} NAD^+ + FADH_2$$

THE CYTOCHROMES

These are haem compounds that are found in all aerobic cells and are important respiratory pigments. They absorb strongly in the visible part of the spectrum and like other haem compounds are red in colour. Oxidation and reduction bring about a shift in the absorption bands and it was this property that led to their discovery by Keilin. Cytochromes are classified a, b, c, etc., by their absorption maxima but those in a particular group are not necessarily related by structure or function.

Cytochromes are important catalysts in cell respiration during which the iron atom is alternately reduced and oxidized. Cytochromes therefore accept and donate electrons rather than hydrogens and several of them may be linked together to form an electron transport chain like that in mitochondria:

$$2\left[\begin{array}{cccccc} & \text{cyt } b^{2+} & \text{cyt } c_1^{3+} & \text{cyt } c^{2+} & \text{cyt } a^{3+} & \text{cyt } a_3^{2+} \\ e^- + \text{cyt } b^{3+} & \text{cyt } c_1^{2+} & \text{cyt } c^{3+} & \text{cyt } a^{2+} & \text{cyt } a_3^{3+} \end{array}\right] \begin{array}{c} 2H^+ + 1/2O_2 \\ \\ H_2O \end{array}$$

7.6   Electron transport: the 'capture' of energy

Energy from oxidations

REDUCTION POTENTIALS

The tendency for an acid (AH) to lose a proton can be described quantitatively in terms of the pK_a value which is a function of the equilibrium constant K_a:

$$AH \rightleftharpoons A^- + H^+$$

$$K_a = [H^+][A^-]/[AH] \quad \text{and} \quad pK_a = -\log_{10} K_a$$

A scale can therefore be established of acid strength and the smaller the pK_a, the stronger the acid.

In a similar way the tendency for a reductant (B) to lose an electron can be defined in terms of the *reduction potential*:

$$B \rightleftharpoons B^+ + e^-$$

The electrode potential of this half-reaction is measured in a complete cell with another electrode to complete the circuit. For reference purposes, reduction potentials are compared to the hydrogen electrode which consists of hydrogen gas bubbled round a platinum wire dipping in a solution of hydrogen ions:

$$H_2 \rightleftharpoons 2H^+ + 2e^-$$

Under standard conditions (temperature = 25 °C, pressure = 1 atmosphere, concentration 1 mol/litre) the *reduction potential E_0* of the hydrogen electrode is assigned a value of zero. *Reductants* therefore have a negative reduction potential since electrons tend to flow to the hydrogen electrode while *oxidants* have a positive value, as electrons are drawn away from the hydrogen electrode.

The standard state of $[H^+] = 1$ mol/litre or pH = 0 is far removed from the conditions in the living cell. For biological systems, standard electrode potentials are measured at pH 7 and given the symbol E_0'. At pH 7, the hydrogen electrode has a potential of -0.42 volts and not zero so redox pairs with values that are more negative than this are reductants and those that are more positive are oxidants.

FREE ENERGY CHANGE

Concentrations in the cell are less than molar and the reduction potential *in vivo*, E', depends on the relative concentrations of the oxidant and reductant in a redox pair:

$$E' = E_0' + RT/nF \ln[\text{oxidant}]/[\text{reductant}]$$

R = gas constant, T = temperature (K), n = number of electrons
F = the Faraday (Avogadro's number × charge on the electron)

The difference in potential between two redox pairs ($\Delta E'$) that are at equilibrium is zero so the change in the standard reduction potential ($\Delta E_0'$) is related to the equilibrium constant K:

$$\Delta E_0' = RT/nF \ln K$$

now

$$\Delta G^{0\prime} = -RT \ln K$$

so that

$$\Delta G^{0\prime} = -nF \Delta E_0'$$

Therefore, the change in free energy between two redox pairs depends on the number of electrons transferred, the Faraday and E_0', the difference in their standard reduction potentials.

Reduction potentials are extremely useful since they enable predictions to be made about the direction of electron flow and the feasibility of a reduction on thermodynamic grounds.

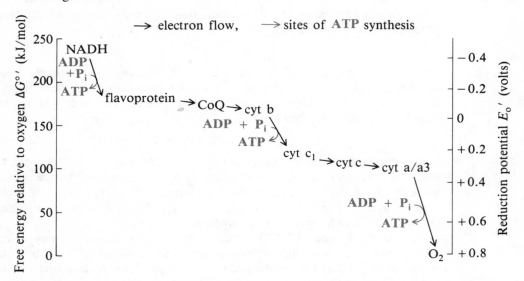

Figure 7.11 The free energy relative to oxygen of the components of the electron transport chain.

ELECTRON TRANSPORT AND THE RELEASE OF ENERGY

Many oxidations take place in the living cell by dehydrogenation, when a pair of hydrogen atoms are transferred from a substrate to an acceptor such as NAD^+. The reduced coenzyme is then oxidized to form NAD^+ and water in the mitochondrial electron transport chain. The energy change of the dehydrogenation step is quite small but a large amount of energy is liberated when the reduced coenzyme is oxidized by molecular oxygen. In the electron transport chain this energy is released in a series of discrete steps, some of which are large enough to provide enough energy to synthesize ATP (Fig. 7.11).

An example of this is given below which shows the energy changes when ethanol is oxidized in the liver cell by alcohol dehydrogenase.

$$\begin{array}{lr} & \Delta G^{0\prime}\ (kJ/mol) \\ CH_3CH_2OH + NAD^+ \rightleftharpoons CH_3CHO + NADH + H^+ & +23 \\ H^+ + NADH + \tfrac{1}{2}O_2 \longrightarrow NAD^+ + H_2O & -220 \\ \hline CH_3CH_2OH + \tfrac{1}{2}O_2 \longrightarrow CH_3CHO + H_2O & -197 \end{array}$$

The electron transport chain

COMPONENTS OF THE RESPIRATORY CHAIN

Redox pairs can be arranged in the order of their reduction potentials and such a series occurs naturally as an electron transport chain on the inner membrane of mitochondria (Fig. 7.12). The four main groups of electron transport carriers in mitochondria are shown below:

1. *NAD*

$$NAD^+ + 2H \rightleftharpoons NADH + H^+$$

2. *Flavoproteins*

$$FMN + 2H \rightleftharpoons FMNH_2$$

$$FAD + 2H \rightleftharpoons FADH_2$$

3. *Coenzyme Q*

$$CoQ + 2H \rightleftharpoons CoQH_2$$

4. *Cytochromes*

$$\text{cytochrome}\ (Fe^{3+}) + e^- \rightleftharpoons \text{cytochrome}\ (Fe^{2+})$$

In mitochondria the chain is present as four membrane-bound complexes and two mobile carriers, coenzyme Q and cytochrome c. Coenzyme Q is hydrophobic and

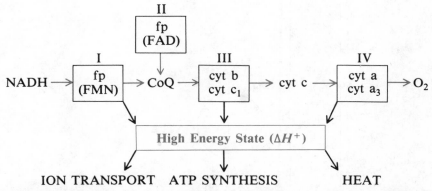

Figure 7.12 The generation and utilization of a high energy state during electron transport (red arrow) in mitochondria (I, II, III, IV = respiratory complexes, fp = flavoproteins, cyt = cytochromes).

moves in a lipid environment while cytochrome c is soluble in aqueous solution. The components are alternately reduced and oxidized as electrons flow down the chain until the hydrogens originally passed on to NAD$^+$ eventually reduce molecular oxygen to water (Fig. 7.12).

OXIDATIVE PHOSPHORYLATION

The synthesis of ATP during electron transport is known as *oxidative phosphorylation* and a great deal of information was accumulated on this topic in many laboratories round the world. There was general agreement that electron flow created a 'high energy state' and that this was then used to drive the synthesis of ATP or other processes (Fig. 7.12) but the nature of this 'high energy state' proved to be very elusive for many years until the advent of the chemiosmotic theory.

CHEMIOSMOTIC THEORY

Several of the reactions of the electron transport chain involve protons and Mitchell suggested that during electron flow hydrogen ions are transferred from the inside to the outside of the mitochondrial inner membrane giving rise to a proton gradient and a potential difference across the membrane:

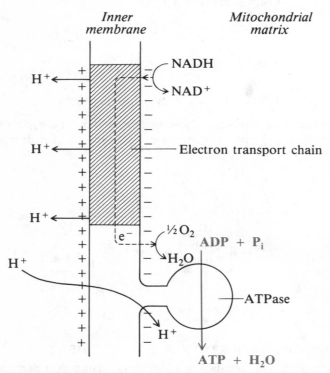

Figure 7.13 Schematic representation of the essential features of the chemiosmotic theory for the generation of ATP.

$$H^+ + NADH + FMN \longrightarrow NAD^+ + FMNH_2$$

$$CoQH_2 + 2cyt\ b(Fe^{3+}) \longrightarrow CoQ + 2cyt\ b(Fe^{2+}) + 2H^+$$

$$2H^+ + 2cyt\ a_3(Fe^{2+}) + \tfrac{1}{2}O_2 \longrightarrow 2cyt\ a_3(Fe^{3+}) + H_2O$$

The mitochondrial inner membrane is impermeable to protons apart from particular sites which contain ATPase. It is here that the hydrogen ions cross the membrane and react with the excess hydroxyl ions in the mitochondrial matrix to form water and drive the ATPase in reverse:

$$ADP + P_i \longrightarrow ATP + H_2O$$

The discharge of the proton gradient and membrane potential is therefore the driving force for the synthesis of ATP (Fig. 7.13).

8. Metabolic fuels

8.1 Glycogen and starch: energy reserves

Glucose is the primary metabolic fuel used by all cells and when there is an excess present in the organism the glucose is polymerized and stored until needed. *Glycogen* is the polymer synthesized in animals; *starch*, which has a similar structure, is laid down in plants. These large molecules are stores of metabolic energy and are broken down to glucose by the organisms when required.

If ATP is the 'energy currency' of the cell, then by the same analogy glucose is the current account at the bank from which the money is drawn and glycogen is the deposit account used to top up the cheque account when funds run low.

Digestion of carbohydrates

BREAKDOWN OF POLYSACCHARIDES

Most carbohydrate in the diet consists of starch which is broken down to glucose in the gastrointestinal tract. The hydrolysis of starch starts in the mouth where *salivary amylase* attacks the α-1,4 bonds of amylose and amylopectin to give maltose and maltotriose—the di- and tri-saccharides of glucose. The digestion continues in the duodenum with *pancreatic amylase* which removes glucose residues until the α-1,6 branch points are reached and a *limit dextrin* is formed. Another enzyme, α-*dextrinase*, then attacks the α-1,6 bonds and the α-1,4 bonds of the limit dextrin to give glucose.

HYDROLYSIS OF DISACCHARIDES

The digestion of starch is completed by the action of *maltase* which catalyses the hydrolysis of maltose to two molecules of glucose (Fig. 8.1).

Two other important sources of glucose in the diet are sucrose from cane or beet and lactose present in milk. Like maltose they are also disaccharides and are hydrolysed to their component monosaccharides by *sucrase* and *lactase* respectively:

$$\text{sucrose} + \text{H}_2\text{O} \xrightarrow{\text{[sucrase]}} \text{D-glucose} + \text{D-fructose}$$

$$\text{lactose} + \text{H}_2\text{O} \xrightarrow{\text{[lactase]}} \text{D-glucose} + \text{D-galactose}$$

Lactase is extremely important in infants but the activity of the enzyme declines during development and may be lost altogether by maturity, so that some adults suffer

amylopectin

↓ [α-amylase]

limit dextrin

↓ [α-1,6-glucosidase]
↓ [α-amylase]

maltose + glucose

↓ [maltase]

glucose

Overall reaction: $(glucose)_n \longrightarrow n$ glucose

Figure 8.1 The digestion of amylopectin (the enzymes are shown in parentheses).

from *lactose intolerance* and become ill on drinking milk. This is particularly common among populations in Africa and the Far East so that sending dried milk to relieve famine in these areas, although helpful for the children, is of little use to the adults. Finally, glucose and the other monosaccharide products of digestion are absorbed by the epithelial cells of the small intestine and are carried in the blood to the liver.

Glycogen metabolism

SYNTHESIS

Glycogen is laid down by many tissues but liver and muscle account for most of the glycogen synthesized in the body (Section 4.1). There are five enzymes involved in the synthesis of glycogen from glucose, a process known as *glycogenesis*.

1. *Hexokinase and glucokinase* The first step involves the phosphorylation of glucose by ATP to form glucose-6-phosphate (G-6-P), a reaction catalysed by *hexokinase* (HK) or *glucokinase* (GK):

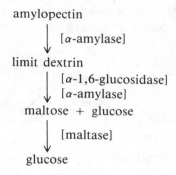

2. *Phosphoglucomutase* The position of the phosphate group is changed by

phosphoglucomutase (PGM) which converts glucose-6-phosphate to glucose-1-phosphate with glucose-1,6-bisphosphate as a catalytic intermediate:

glucose-6-phosphate   glucose-1-phosphate

3. *UDP-glucose pyrophosphorylase* Glucose-1-phosphate is then 'activated' with uridine triphosphate (UTP) to give UDP-glucose and pyrophosphate, a reaction catalysed by the enzyme *UDP-glucose pyrophosphorylase* (UDPG-Pyr). An active *pyrophosphatase* (Pyr) splits the pyrophosphate into two molecules of orthophosphate making the reaction effectively one way:

$$\text{glucose-1-phosphate} + \text{UTP} \underset{\text{[UDPG-Pyr]}}{\rightleftharpoons} \text{UDP-glucose} + \text{PP}_i$$

$$\text{PP}_i \xrightarrow{\text{[Pyr]}} \text{P}_i + \text{P}_i$$

4. *Glycogen synthase* The UDP-glucose now donates glucose residues to an existing glycogen molecule creating new α-1,4 links and extending the straight chain parts of the polymer; a reaction catalysed by *glycogen synthase* (GS):

$$\text{UDP-glucose} + (\text{glycogen})_n \xrightarrow[\text{Mg}^{2+}]{\text{[GS]}} \text{UDP} + (\text{glycogen})_{n+1}$$

5. *Branching enzyme* The addition of new glucose residues continues until the straight chain is 11 or more segments long. Then a section of chain of about seven glucose units is removed to a more internal site and a new α-1,6 branch point is created by branching enzyme (BM).

BREAKDOWN
The conversion of glycogen to glucose, known as *glycogenolysis*, takes place in the liver when the blood glucose falls and, although many of the intermediates are the same as in *glycogenesis*, the breakdown of glycogen follows a different pathway to that of its synthesis (Fig. 8.2).

1. *Phosphorylase and debranching enzyme* The enzyme phosphorylase [Ph] catalyses the sequential removal of glucose units from the non-reducing end of the polymer to form glucose-1-phosphate (G-1-P):

$$(\text{glucose})_n + \text{P}_i \xrightarrow{\text{[Ph]}} \text{glucose-1-phosphate} + (\text{glucose})_{n-1}$$

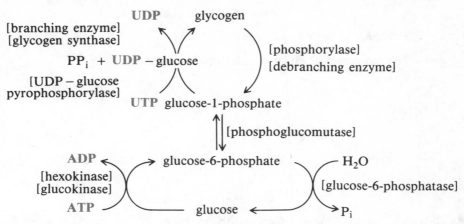

GLYCOGENESIS   *GLYCOGENOLYSIS*

Figure 8.2 The synthesis and breakdown of liver glycogen. The reactions are the same in muscle apart from the absence of the enzyme glucose-6-phosphatase.

The reaction is freely reversible *in vitro* but the high ratio of P_1 to G-1-P in the cell means that the formation of glucose-1-phosphate is irreversible *in vivo*. This degradation continues until four glucose residues from a branch point are reached. *Debranching enzyme* then removes three of the glucosyl residues leaving the α-1,6 residue of the branch point. It transfers the triglucose unit to a main branch of the polymer with the formation of an α-1,4 linkage susceptible to phosphorylase action. The enzyme then removes the α-1,6 glucosyl residue by simple hydrolysis to yield glucose. Thus the whole glycogen molecule can be mobilized by the combined actions of phosphorylase and debranching enzyme (Fig. 8.3).

2. *Phosphoglucomutase*   This enzyme is freely reversible and catalyses the same reaction as that found in glycogenesis but in the opposite direction so that G-1-P is converted to G-6-P.

glucose-6-phosphate   glucose

3. *Glucose-6-phosphatase*   This enzyme is active in the liver but is not present in muscle, which is consistent with the different roles of liver and muscle glycogen. The glycogen in the liver is a reservoir of glucose for the whole body and the liver must be

○ glucose residues in core

○ glucose residues in side chain

Phosphorylase

Ten molecules of glucose-1-phosphate released.

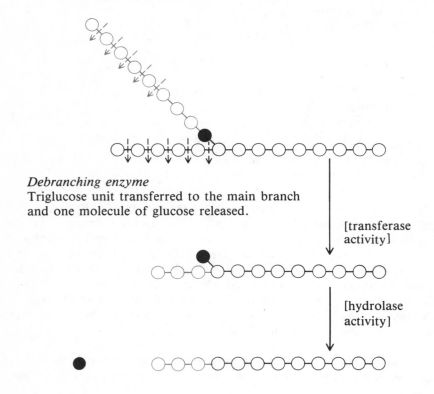

Debranching enzyme
Triglucose unit transferred to the main branch
and one molecule of glucose released.

[transferase activity]

[hydrolase activity]

The phosphorylase continues until degradation is complete.

Figure 8.3   The breakdown of glycogen by phosphorylase and debranching enzyme.

able to hydrolyse G-6-P to glucose which is then carried by the blood to all the tissues. The glycogen in muscle, on the other hand, is a store of energy which is used in that tissue and does not contribute to the blood glucose.

A summary of the reactions and enzymes involved in the breakdown of glycogen is shown in Fig. 8.2.

GLYCOGEN STORAGE DISEASES
These are a group of rare inherited diseases known as the *glycogenoses* in which the

Table 8.1 Some examples of glycogen storage disease

Type	Name	Defective enzyme	Glycogen Amount	Glycogen Structure	Organs affected	Clinical signs
I	Von Gierke's disease	glucose-6-phosphatase	↑	normal	liver kidney intestine	large liver severe hypoglycaemia failure to thrive
II	Pompe's disease	α-glucosidase (in lysosomes)	↑↑	normal	all organs	cardiac failure death in infancy
III	Forbe's limit dextrinosis	amylo-1,6-glucosidase	↑	short outer branches	liver muscle	like Type I but milder
IV	Anderson's amylopectinosis	branching enzyme	—	long outer branches	all organs	progressive cirrhosis liver failure death in infancy
V	McArdle's disease	phosphorylase	↑	normal	muscle	muscle cramps fatigue

tissues contain large quantities of glycogen or glycogen with an unusual structure. They are all caused by a missing or defective enzyme although the enzyme defect is not always manifested in all organs. The severity of the disease depends on the enzyme which is missing and many of the glycogenoses are fatal. Some examples of glycogen storage diseases are given in Table 8.1.

Starch in plants

STORAGE

The synthesis of starch in plants is similar in many ways to that of glycogen in animals with some minor differences. As with glycogen, the precursor for starch is glucose-1-phosphate which is activated by *ADP-glucose pyrophosphorylase* (ADPG-Pyr):

$$\text{ATP} + \text{glucose-1-phosphate} \xrightarrow{\text{[ADPG-Pyr]}} \text{ADP-glucose} + \text{PP}_i$$

This reaction, although reversible in the test tube, is irreversible in the cell because of an active pyrophosphatase which rapidly removes pyrophosphate. The pyrophosphorylase is stimulated by 3-phosphoglycerate produced by photosynthesis so that the carbon fixed is used to extend the existing chain length of starch in the granules. *Starch synthase* (SS) then catalyses the synthesis of α-1,4 links in amylose and amylopectin from ADP-glucose and a starch primer:

$$(\text{starch})_n + \text{ADP-glucose} \xrightarrow{\text{[SS]}} (\text{starch})_{n+1}\ \text{ADP}$$

α-1,6 branch points are introduced by branching enzymes: *Q enzyme* converts the straight-chain amylose to the branched amylopectin and *amylopectin branching enzyme* introduces more branch points into the amylopectin.

Starch in plants, like glycogen in animals, is stored in the cell as water-insoluble granules the shape and size of which depend on the particular plant, for example in potatoes most granules have a diameter from 50 to 100 μm. Tubers such as potatoes and seeds such as corn are rich stores of starch and provide a staple food for man.

UTILIZATION

The enzymes involved in the breakdown of glycogen all have their parallels in the degradation of starch. However, phosphorylase does not attack starch granules unless they have been partially degraded by hydrolytic enzymes. α-Amylase is important in the breakdown of starch in brewing and plants also contain a β-amylase which hydrolyses the 1,4 links of starch to give β-maltose. Both amylases appear to play a role in the rapid mobilization of starch during germination.

8.2   Glucose: its catabolism in glycolysis

The role of glucose in metabolism

BLOOD GLUCOSE

Glucose is one of the two major fuels of the body and is an important source of energy for almost all forms of life. It occupies a central role in cellular metabolism and in animals the concentration in the blood is carefully controlled to ensure that there is an adequate supply of this metabolic fuel to all parts of the body. Most tissues are able to use fatty acids as an alternative fuel but the brain is dependent on glucose and any drastic fall in the blood level leads to coma and eventually death.

SOURCE OF GLUCOSE

In plants glucose comes from carbohydrates manufactured during photosynthesis; in animals the main source of glucose is starch and other carbohydrate stores of plants. In humans the carbohydrates of cereals are of major economic importance since they provide a source of energy that is considerably cheaper than the protein and fat of meat.

In animals blood glucose comes from ingested carbohydrate either directly by digestion and absorption of food or indirectly from glucose stored as its polymer glycogen in the liver. During starvation, when dietary carbohydrate is unavailable and glycogen stores are depleted, glucose is synthesized from non-carbohydrate starting materials such as amino acids in *gluconeogenesis*.

Glycolysis

THE NATURE OF GLYCOLYSIS

Glucose enters the tissues of the body where it is oxidized and the metabolic energy released and captured. This oxidation of glucose in the cell involves two metabolic pathways: *glycolysis* and the *Krebs' cycle* (Fig. 8.4).

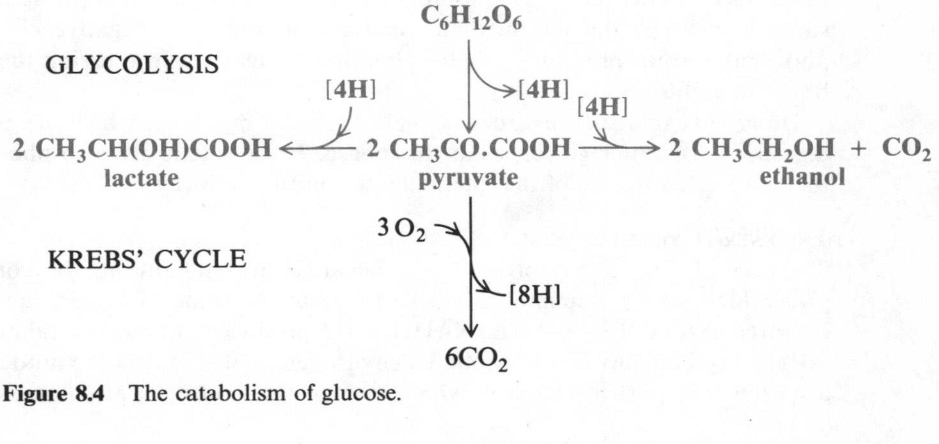

$$C_6H_{12}O_6$$

GLYCOLYSIS

[4H]   [4H]   [4H]

2 $CH_3CH(OH)COOH$ $\longleftarrow$ 2 $CH_3CO.COOH$ $\longrightarrow$ 2 CH_3CH_2OH + CO_2
lactate   pyruvate   ethanol

KREBS' CYCLE

$3 O_2$

[8H]

$6CO_2$

Figure 8.4   The catabolism of glucose.

Glycolysis takes place in ten steps and the reader meeting the metabolic pathway for the first time may be quite bewildered by the long list of substrates, enzymes and cofactors that are involved. However, before learning all the names, it is vital to get an overall view of what happens in glycolysis and this is how all metabolic pathways should first be approached.

Glycolysis comes from the Greek words *glycos* meaning sugar and *lysis* meaning dissolution and basically involves the splitting of the six-carbon glucose into two three-carbon fragments. At the same time two pairs of hydrogen atoms are removed and transferred to NAD^+ to give $NADH + H^+$. Glycolysis can only continue in the cell if the reduced coenzyme ($NADH + H^+$) is reoxidized and it completes its catalytic role. This can be achieved in one of two ways:

1. Reoxidation in mitochondria under aerobic conditions in cells that possess mitochondria.
2. Reoxidation in the cytosol under anaerobic conditions by reduction of an acceptor
   (a) pyruvate $\longrightarrow$ lactate
   (b) pyruvate $\longrightarrow$ acetaldehyde $\longrightarrow$ ethanol

This anaerobic glycolysis is common in micro-organisms but also occurs in rapidly contracting muscle and in cells such as erythrocytes that do not possess mitochondria.

The oxidation of glucose is then completed in the Krebs' cycle by steps involving several dehydrogenations and the formation of carbon dioxide. The hydrogens removed are then oxidized to water in the electron transport chain with the formation of ATP.

THE ENERGY-UTILIZING PHASE

In the first part of glycolysis, glucose is converted to two molecules of glyceraldehyde-3-phosphate (Gly-3-P) via a series of phosphorylated intermediates (Fig. 8.5). This phase has a positive ΔG and is driven by the hydrolysis of ATP so that phosphate activates the sugars by raising them to a higher energy level prior to breakdown, rather like a match being used to light a fire. The introduction of phosphate into the sugar also keeps the intermediates in the cell, since the negative charge on the phosphate group prevents the phosphorylated sugars from crossing the cell membrane by diffusion.

Other D sugars such as fructose, galactose and mannose, which are produced in digestion, also enter glycolysis at this phase. These sugars are also phosphorylated then converted to one of the intermediates on the pathway.

THE ENERGY-YIELDING PHASE

The second part of glycolysis has a negative ΔG and during the conversion of glyceraldehyde-3-phosphate (Gly-3-P) to pyruvate, some of the energy released is captured as ATP (Fig. 8.6). The $NADH + H^+$ produced during the oxidation of (Gly-3-P) by glyceraldehyde-3-phosphate dehydrogenase then enters the mitochondria by a special transport mechanism, where each pair of hydrogen atoms are oxidized to

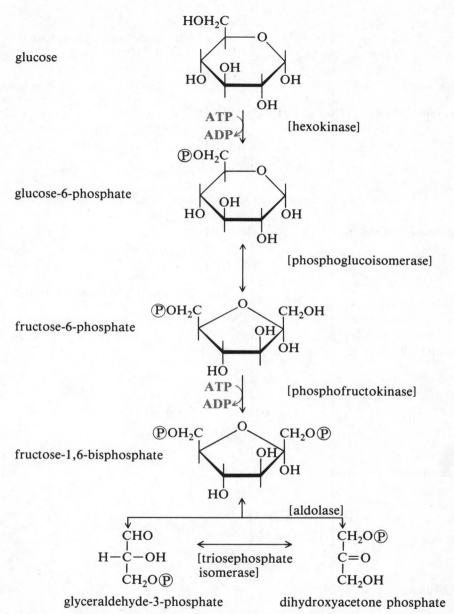

Figure 8.5   The activation of glucose in glycolysis (the enzymes are shown in square brackets).

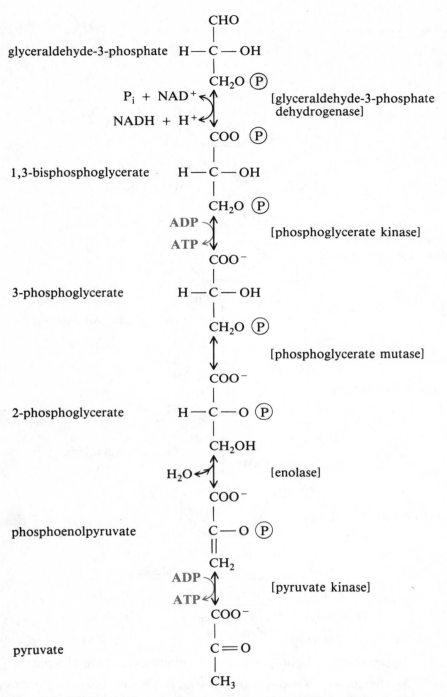

Figure 8.6 The production of adenosine triphosphate (ATP) in glycolysis (the enzymes are shown in square brackets).

water. At the same time ADP is phosphorylated to give ATP and the whole process is given the highly descriptive name *oxidative phosphorylation*. The steps involving the enzymes phosphoglycerate kinase and pyruvate kinase also give rise to ATP but this time without the consumption of oxygen and this process is known as *substrate level phosphorylation*.

CONTROL POINTS

The rate of glycolysis depends on a number of metabolites of which the blood glucose and the cellular ATP are probably the most important. The concentration of these compounds affects the activity of several enzymes including the following:

Hexokinase   This enzyme is very active in brain, muscle and other tissues and has a K_m for glucose of about 0.1 mM. The blood glucose is much greater than this, by an order of magnitude, so that changes in the concentration of glucose in the blood will have no effect on the activity of *hexokinase* (HK). This is only true of course if the cell membrane is freely permeable to glucose and this is the case for brain. Hexokinase therefore ensures that glucose continues to be used even when the blood glucose is low and this is extremely important since the brain can normally only use glucose as a metabolic fuel.

Hexokinase is also inhibited by its product glucose-6-phosphate (G-6-P), so that when the concentration of G-6-P increases as glycolysis slows down, the rate at which glucose is phosphorylated falls.

Glucokinase   This enzyme, which is found in the liver, has a K_m for glucose of about 10 mM and, since the concentration of glucose in the blood varies from 4 to 8 mM, this means that *glucokinase* (GK) is sensitive to changes in the blood glucose. A high blood glucose speeds up the rate of phosphorylation, while a low blood glucose slows down the rate at which G-6-P is formed. Furthermore, glucokinase is not inhibited by G-6-P so that when the concentration of this compound rises, following a reduction in the rate of glycolysis, phosphorylation of glucose continues and the glucose-6-phosphate is converted to glycogen. This is consistent with the role of the liver which stores glucose when the blood sugar is high and breaks down glycogen when this falls.

Phosphofructokinase   This is an extremely important control point and *phosphofructokinase* (PFK) is inhibited by high concentrations of ATP in tissues such as muscle. This inhibition is relieved by increases in the concentration of AMP so the rate of glycolysis in muscle depends to a large extent on the ratio of ATP to AMP.

Pyruvate kinase   High concentrations of ATP also inhibit *pyruvate kinase* (PK) of liver which slows down the rate of glycolysis in this organ.

ATP level   This means that glycolysis is stimulated when the ATP is low and the cell needs energy and slowed down when there is enough ATP available.

Energetics of glycolysis

GLUCOSE TO PYRUVATE

Two molecules of glyceraldehyde-3-phosphate are formed from one molecule of glucose so that on the credit side 2×3 molecules of ATP are obtained by oxidative phosphorylation and 2×2 molecules of ATP by substrate level phosphorylation making 10 in all. However, on the debit side, two molecules of ATP are required to activate one molecule of glucose so that the net yield is eight molecules of ATP under aerobic conditions.

FERMENTATION

The formation of pyruvate from glucose requires oxygen for oxidative phosphorylation so it is wrong to talk about glycolysis as an *anaerobic pathway*. There is however an anaerobic option for glycolysis when the $NADH + H^+$, formed during oxidation of Gly-3-P, is oxidized to NAD^+ during the conversion of pyruvate to ethanol or lactate (Fig. 8.4). Ethanol therefore is actually a waste product of yeast and other micro-organisms. This is then drunk as wine, beer or spirits by human beings so just think how dependent human society is on the excretory product of a micro-organism.

Lactate is formed in muscle when the oxygen supply is insufficient to oxidize all the pyruvate in the Krebs' cycle so that some of the pyruvate is converted to lactate.

The anaerobic option is energetically inefficient since no ATP is formed by oxidative phosphorylation so there is a net gain of only two ATPs per molecule of glucose by substrate level phosphorylation. The energy for the formation of the two molecules of ATP comes from that released during the oxidation of an aldehyde (glyceraldehyde-3-phosphate) to an acid (1,3-bisphosphoglycerate). The ATP yield for fermentation is low, yet many micro-organisms grow quite well under anaerobic conditions and generate enough ATP for their requirements. They do this by speeding up the rate at which glucose is metabolized when oxygen is absent; a phenomenon known as the *Pasteur effect* after its discoverer Louis Pasteur.

SUMMARY

Glycolysis is simply a device for cleaving glucose to two molecules of triosephosphates, oxidizing an aldehyde to an acid and conserving some of the energy released as ATP.

Glycolysis also serves as a source of pyruvate, a key intermediate in the metabolism of fats and amino acids, and this will be discussed later.

8.3   Triglyceride: storage and mobilization

Fat, in the form of triglyceride, is the store for fatty acids the other metabolic fuel used by living organisms. Ingested fat is broken down and oxidized and any dietary excess is stored in special tissues as a long-term reserve of metabolic energy. In the plant kingdom, some seeds contain large quantities of triglyceride and these stores of oil are the main source of energy during germination, until the young seedling is capable of photosynthesis. In animals, fat is stored in adipose tissue and mobilized when other sources of metabolic energy such as blood glucose and liver glycogen are low.

When energy is needed by the organism, triglycerides from the diet or in the fat stores are hydrolysed and the resulting glycerol and fatty acids are then oxidized to carbon dioxide and water with the generation of ATP. The storage and mobilization of fat in animals is dealt with in this section, while the oxidation of fatty acids is discussed in Section 8.4.

Digestion

HYDROLYSIS OF TRIGLYCERIDES

Fats are insoluble in water and dietary fats are emulsified in the gastrointestinal tract by the detergent action of bile salts before being hydrolysed by *pancreatic lipase*. The digestion is about 95 per cent complete and the main products are 2-monoacyl glycerol and the fatty acids from positions 1 and 3:

$$_2HCO \cdot COR_1 \qquad\qquad\qquad _2HCOH$$
$$| \qquad\qquad\qquad\qquad\qquad |$$
$$R_2OC \cdot OCH + H_2O \xrightarrow[\substack{[\text{bile} \\ \text{salts}]}]{[\text{lipase}]} R_2OC \cdot OCH + R_1COO^- + R_3COO^- + 2\,H^+$$
$$| \qquad\qquad\qquad\qquad\qquad |$$
$$_2HCO \cdot COR_3 \qquad\qquad\qquad _2HCOH$$

ABSORPTION AND RE-ESTERIFICATION

The sodium salts of the fatty acids and the 2-monoacylglycerol form mixed micelles with bile salts and the products of hydrolysis enter the intestinal mucosal cells by passive diffusion. The absorption is quite rapid as the concentrations of free fatty acids and 2-monoacylglycerol in the mucosal cells are low because of the rapid resynthesis of the triglycerides. The long-chain fatty acids that enter the mucosal cells are converted to the corresponding fattyacyl-CoAs by the action of *acyl CoA synthetase* (ACS).

$$R_4COOH + ATP + CoA\text{-}SH \xrightarrow[\text{Mg}^{2+},\,\text{K}^+]{[\text{ACS}]} R_4CO\text{—}S\text{—}CoA + AMP + PP_i$$

These fattyacyl CoAs then react with the 2-monoacylglycerol to give diacylglycerol then triacylglycerol under the influence of *monoacylglycerol transacylase* (MT) and *diacylglycerol transacylase* (DT):

$$_2HCOH \quad R_4CO\text{—}S\text{—}CoA \qquad\qquad _2HCO \cdot COR_4$$
$$| \qquad\qquad\qquad\qquad\qquad\qquad\qquad |$$
$$R_2OC \cdot OCH \;+\; \xrightarrow[\text{[DT]}]{\text{[MT]}} R_2OCO \cdot CH \quad + 2\,CoA\text{-}SH$$
$$| \qquad\qquad\qquad\qquad\qquad\qquad\qquad |$$
$$_2HCOH \quad R_5CO\text{—}S\text{—}CoA \qquad\qquad _2HCO \cdot COR_5$$

At first sight, the breakdown and resynthesis of triglycerides appears to be a pointless exercise and a wasteful expenditure of energy but closer examination shows that this is not the case for the following reasons:

1. The intact triglycerides do not diffuse through the mucosal cell wall very efficiently compared with the hydrolysis products from the mixed micelles.
2. This process enables the fatty acid composition of the triglycerides to be controlled so that the fats synthesized are characteristic of the animal and not the fat composition of the diet.
3. The direct absorption of fatty acids by the cell would lead to acidosis.
4. Triglyceride can be transported as chylomicrons in the blood without risk of metabolism.

This last point is very important since only those tissues possessing an active lipoprotein lipase can use the triglyceride and in the fed animal this is in only white adipose tissue. The resynthesis therefore is an essential part of the mechanism whereby dietary triglyceride is diverted directly into storage and is not oxidized.

An adequate supply of bile salts is needed for the absorption of fats and the importance of these natural detergents can be seen in disease states where bile is absent from the gut. Under normal circumstances only about 5 per cent of the dietary fat appears in the stool but, in the absence of bile salts, this rises to 50 per cent or more, a condition known as *steatorrhoea*.

TRANSPORT

The shorter chain fatty acids below C_{10} are not esterified and are sufficiently soluble to be carried in the blood plasma to the liver where they are metabolized. Most of the fat is present as intact triacylglycerol and this is incorporated into emulsified droplets known as *chylomicrons* (Fig. 8.7) which are absorbed by the lymphatic system and pass into the blood via the thoracic duct. The presence of tiny fat droplets in the blood gives rise to the characteristic 'milky appearance' of blood plasma after a fatty meal.

Composition %

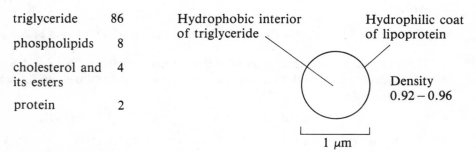

triglyceride	86
phospholipids	8
cholesterol and its esters	4
protein	2

Hydrophobic interior of triglyceride

Hydrophilic coat of lipoprotein

Density 0.92 – 0.96

1 μm

Figure 8.7 The composition of chylomicrons.

Adipose tissue

DEPOSITION OF TRIGLYCERIDE

The triglyceride in the chylomicrons is stored in specialized cells of connective tissue known as adipocytes and the fat is present as a large droplet in the cytoplasm which can occupy as much as 90 per cent of the cell volume.

An adipocyte

- Membrane
- Nucleus
- Fat droplet of triglyceride (TG)

The uptake of triglyceride into adipocytes is similar to the uptake of fat by the mucosal cells of the gut. A lipoprotein lipase on the internal capillary wall in adipose tissue hydrolyses the triglyceride of the chylomicrons and the resulting glycerol and free fatty acids pass into the fat cells and are re-esterified to triglyceride. In the fed state, white adipose tissue is the only tissue with significant amounts of lipoprotein lipase although other tissues (muscle, heart, lung, etc.) possess an active enzyme under some physiological conditions.

WHITE ADIPOSE TISSUE

Most fat is stored in white adipose tissue and pads of this material are distributed throughout the body where they protect organs and joints against mechanical shock, act as a heat insulator and provide a long-term store of food. The distribution of fat in the body is clearly different in the two sexes and is the reason for the characteristic shapes of the mature male and female form.

BROWN ADIPOSE TISSUE

This type is less common than white adipose tissue and is found in only a few areas of the body. The brown colour is due to the high concentration of cytochromes that are

glycerol   L-glycerol-   dihydroxyacetone   pyruvate
                  3-phosphate   phosphate

Figure 8.8 The metabolism of glycerol in the liver (GLK = glycerol kinase, GPDH = glycerol phosphate dehydrogenase).

present in the numerous mitochondria which appear to be largely uncoupled in the stimulated state. This means that most of the metabolic energy in the tissue is released as heat instead of being captured as ATP. Brown adipose tissue therefore has a high temperature and can be regarded as a 'local heating furnace'. It is particularly prevalent in neonates, which tend to lose heat rapidly, and hibernating animals, which need to generate heat while having a low metabolic rate. Some biochemists have suggested that the reason why some people gain weight, while others on an identical

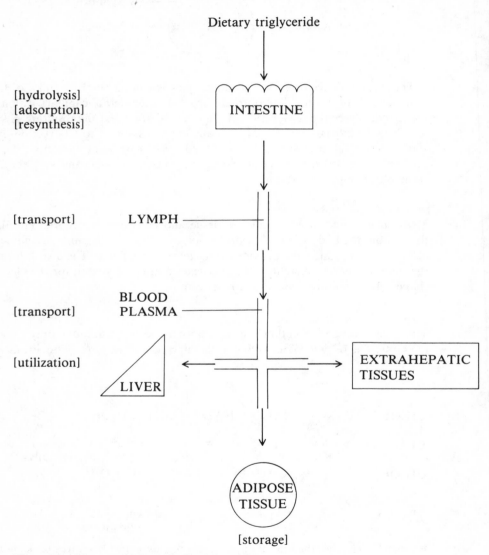

Figure 8.9 The metabolism of triacylglycerols.

diet do not, depends on the amount of brown fat that has survived from childhood. The greater the amount of brown fat, the more unwanted kilojoules can be removed as heat instead of being deposited as fat.

MOBILIZATION OF TRIGLYCERIDE

Fat is mobilized when other sources of energy are depleted. Under these conditions adrenaline and glucagon stimulate a *hormone-sensitive lipase* which is present inside the adipocytes and catalyses the hydrolysis of triacylglycerols to fatty acids and glycerol. The fatty acids leave the cell by passive diffusion and are transported in the blood bound to albumin. From here they enter the tissues by diffusion and the metabolic energy is liberated by β-oxidation and the Krebs' cycle.

GLYCEROL

The other product of hydrolysis is water soluble and is carried to the liver where it is metabolized to dihydroxyacetone phosphate, an intermediate of glycolysis (Fig. 8.8). Further catabolism is possible via glycolysis and the Krebs' cycle but glycerol is most likely to be converted to glucose by gluconeogenesis. This is because the liver would be in a gluconeogenic state under conditions involving the mobilization of triglyceride (e.g. starvation).

The relationship between dietary lipid, fat stores and the use of triglyceride in tissues is summarized in Fig. 8.9.

8.4 Fatty acids: breakdown by β-oxidation

The oxidation of fatty acids

FATTY ACIDS AS A METABOLIC FUEL

Fatty acids are a particularly rich source of potential metabolic energy since the carbon is in a more reduced state than in carbohydrates (Section 3.3). They can be oxidized by most tissues apart from brain and mammalian erythrocytes which can normally only use glucose.

As with glucose, the oxidation of fatty acids takes place in two stages. The first is the breakdown to acetyl-CoA by β-oxidation and the second is the oxidation of the acetyl-CoA to carbon dixoide and water in the Krebs' cycle (Section 8.5).

EARLY EXPERIMENTS

The first indication of β-oxidation was obtained by Knoop in 1909 who fed dogs with fatty acids labelled with a benzene ring on the terminal (ω) carbon atom. He collected the urine and showed that fatty acids with an even number of carbon atoms gave phenylacetate as the end product, while those with an odd number of carbon atoms yielded benzoate.

phenyl acetate   benzoate

He concluded from these results that fatty acids are degraded by the removal of two carbon atoms at a time, a process known as β-oxidation. Knoop could only look at the whole animal in terms of what is fed and the final excretion product but the advent of isotopes and viable tissue preparations enabled many biochemists, including Lynen and Lehninger, to discover the detailed chemical changes and the subcellular localization of this process.

Preparing for oxidation

ACTIVATION OF FATTY ACIDS

The first stage in the metabolism of fatty acids is their reaction with ATP and coenzyme A (CoA—SH) to form fattyacyl-CoA (R—CO—S—CoA). This reaction is catalysed by *acyl-CoA synthetase* present in the outer mitochondrial membrane (Fig. 8.10) and takes place in two stages:

1. $RCOOH + ATP \rightleftharpoons R—CO—AMP + PP_i$

2. $R—CO—AMP + HS—CoA \rightleftharpoons R—CO—S—CoA + AMP$

The free energy change of the overall reaction is only small so the reaction is freely reversible *in vitro*. However, the reaction *in vivo* is essentially one way because of a pyrophosphatase which catalyses the hydrolysis of pyrophosphate to two molecules of orthophosphate. This removes one of the products and so pulls the reaction towards the synthesis rather than the breakdown of fattyacyl-CoA. The first step can be regarded as a priming reaction with the hydrolysis of ATP providing the energy for the formation of the thioester bond of the fattyacyl-CoA.

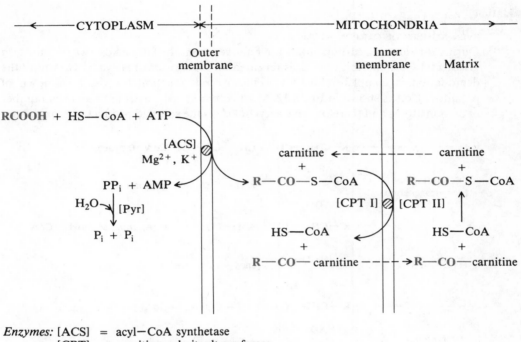

Enzymes: [ACS] = acyl–CoA synthetase
[CPT] = carnitine palmitoyltransferase
[Pyr] = pyrophosphatase
----- = Diffusion

Figure 8.10 The activation and transport of fatty acids into the mitochondrion.

TRANSPORT INTO MITOCHONDRIA

The long-chain fattyacyl-CoAs are very large molecules and do not readily enter mitochondria but are transported through the inner membrane as fattyacyl-carnitine, then reformed within the mitochondria (Fig. 8.10). This is an important control point in the metabolism of fatty acids and, when there is more glucose than can be stored as glycogen, this is converted to acetyl-CoA which is the starting point for the synthesis of fatty acids. An early metabolite on the synthetic pathway is malonyl-CoA and this inhibits carnitine acyltransferase and stops β-oxidation by preventing the uptake of fatty acids into the mitochondria.

$$(CH_3)_3N^+ \text{—} CH_2 \text{—} CH \text{—} CH_2 \text{—} COO^-$$

with a side chain:

$$\underset{R}{\underset{|}{C}} = O \qquad \text{fatty acyl-carnitine}$$

(O connecting to C=O, R below)

β-Oxidation

EVEN NUMBER OF CARBON ATOMS

During β-oxidation, carbon atoms are removed from the fatty acids, two at a time, to give acetyl-CoA and this process is repeated on the shortened fattyacyl-CoA until the degradation is complete (Fig. 8.11). The overall reaction for the breakdown of palmitoyl-CoA is shown in Fig. 8.12. Most common fatty acids have an even number of carbon atoms and therefore give acetyl-CoA as the sole end product of β-oxidation.

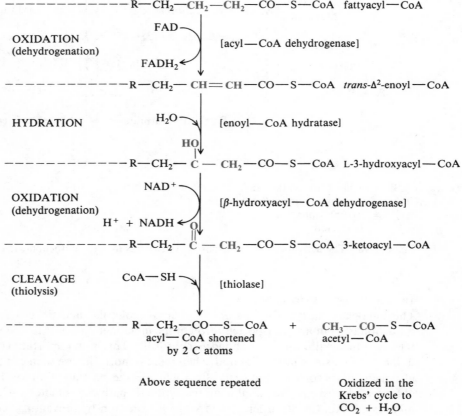

Figure 8.11 The β-oxidation of fatty acids.

$$CH_3(CH_2)_{14}CO—S—CoA + 7CoA—SH + 7FAD + 7NAD^+ + 7H_2O$$

$$\downarrow$$

$$8CH_3CO—S—CoA + 7FADH_2 + 7NADH + 7H^+$$

Fate of products	oxidized in the Krebs' cycle to $CO_2 + H_2O$	CoQ electron transport chain	FMN
Molecules of ATP	96 +	14 +	21 = 131

Figure 8.12 The β-oxidation of palmitoyl-CoA.

ODD NUMBER OF CARBON ATOMS

Fatty acids with an odd number of carbon atoms are rare in most mammalian tissues although they are important in ruminants. The oxidation takes place in exactly the same way as with fatty acids possessing an even number of carbon atoms, with the removal of two carbons at a time to give acetyl-CoA but the last sequence of β-oxidation gives propionyl-CoA rather than acetyl-CoA. The propionyl-CoA is then converted to succinyl-CoA, an intermediate of the Krebs' cycle. Therefore, propionyl-CoA is exceptional in that it can be converted to glucose and the ability to do this is particularly important in ruminants.

UNSATURATED FATTY ACIDS

In the case of unsaturated fatty acids, the metabolism proceeds as normal until a double bond is reached.

cis-Δ^2-*Fattyacyl-CoA* If the *cis* double bond of the product is from C-2 to C-3, the enoylhydratase gives D-hydroxyacyl-CoA and not the L isomer which comes from the *trans*-Δ^2-fattyacyl-CoA (Fig. 8.11). The D isomer is converted to L-hydroxyacyl-CoA by an *epimerase* and the process continues as normal.

cis-Δ^3-*Fattyacyl-CoA* If the *cis* double bond is from C-3 to C-4 of the product, then an isomerase converts this to the *trans*-Δ^2-fattyacyl-CoA and the breakdown then continues as before.

Ketone bodies

FORMATION AND BREAKDOWN

Acetyl-CoA is normally oxidized to carbon dioxide and water in the Krebs' cycle but if there is an increase in the breakdown of fatty acids and a relative lack of

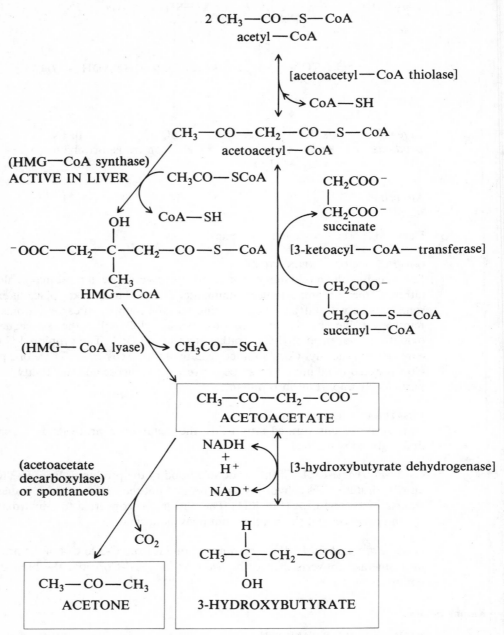

Figure 8.13 The formation and breakdown of ketone bodies, indicated by red boxes (HMG-CoA = 3-hydroxy-3-methyl-glutaryl-coenzyme A).

oxaloacetate from carbohydrate metabolism, then the excess acetyl-CoA is converted to ketone bodies (Fig. 8.13). These compounds pass into the blood from the liver and are metabolized by other tissues such as heart, muscle and kidney cortex where they are converted back to acetyl-CoA and oxidized in the Krebs' cycle (Fig. 8.13). However, liver is unable to do this so if the rate of formation of ketone bodies by the liver is greater than their metabolism by extrahepatic tissues, then the level of these compounds rises in the blood.

KETOSIS

This condition is known as ketosis and if severe gives rise to *acidosis* and all the problems associated with this condition.

Starvation   Ketosis is found in starvation when the glycogen stores have been depleted and fat is the dominant metabolic fuel. Even an overnight fast produces a mild ketosis and this is particularly noticeable in infants when acetone can be detected in their breath first thing in the morning.

Ketotic cows   Ketosis can occur in cows especially when they have just given birth. Cows are particularly dependent on propionate produced by bacteria in the rumen and gluconeogenesis for the normal supplies of glucose. However, during lactation there is a heavy demand on carbohydrate for lactose synthesis and also triglyceride for the synthesis of milk fats and as a source of energy. This in turn leads to a fall in oxaloacetate needed to oxidize acetyl-CoA which is therefore diverted to ketone bodies.

Diabetes   Ketosis also happens in insulin-dependent diabetes when the balance between carbohydrate and fat metabolism is disturbed because of the lack of insulin. The high circulating levels of glucose in the blood and its loss in the urine means that cells are deprived of carbohydrate and have to rely on fat as their main source of energy. This is further exacerbated by the loss of insulin inhibition of hormone-sensitive lipase which leads to an excessive release of fatty acids and ketosis. Further details on diabetic ketosis and other metabolic disturbances caused by diabetes are discussed in Section 12.4.

8.5   Acetyl coenzyme A: oxidation in the Krebs' cycle

Acetyl coenzyme A and the metabolism of carbohydrate and fat

The oxidation of the metabolic fuels discussed so far is incomplete. Glucose from ingested food or stored carbohydrate is split into two molecules of pyruvate in glycolysis, while fatty acids from dietary or stored triglyceride are broken down only as far as acetyl coenzyme A (acetyl-CoA) during β-oxidation. Pyruvate can be converted to acetyl-CoA by decarboxylation and after this the oxidation of carbohydrate and fat follows a common pathway with acetyl-CoA as the meeting point.

OXIDATION OF ACETYL COENZYME A

The oxidation of glucose and fatty acids is completed when acetyl-CoA is catabolized to carbon dioxide and water in the Krebs' cycle. The overall effect is that the two carbon atoms of the acetyl group are oxidized to two molecules of carbon dioxide and four pairs of hydrogen atoms are removed and passed on to the electron transport chain with the formation of ATP. The equation for this can be written:

$$CH_3CO—S—CoA + 3H_2O \longrightarrow 2CO_2 + 4 \times 2[H] + CoA\text{-}SH$$

The actual metabolic sequence is cyclic and not linear and the two carbons of the acetyl group combine with a four-carbon compound to give a six-carbon compound which then loses carbon dioxide in two stages to form a four-carbon compound. This then undergoes a series of changes to form the first four-carbon intermediate:

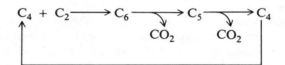

DISCOVERY OF THE CYCLE

Prior to the Second World War a group of workers under Szent-Gyorgi had shown that the respiration of minced pigeon breast muscle was stimulated by small quantities of succinate, fumarate, malate and oxaloacetate. The increase in oxygen uptake was greater than could be accounted for by the oxidation of the substrates themselves, suggesting that they acted catalytically on respiration. Krebs extended this observation and showed that the five-carbon 2-oxoglutarate and the six-carbon citrate also stimulated respiration in this way. Other biochemists had by then shown that tissues contained enzymes which acted on these compounds so that they could be related in a metabolic sequence. Krebs, while studying the oxidation of pyruvate, showed that all of these carboxylic acid salts stimulated the oxidation of pyruvate. He also showed that malonate, which inhibited the oxidation of succinate to fumarate, blocked the oxidation of pyruvate and caused a build-up of succinate whatever the carboxylate salt used to stimulate respiration. From this and other evidence Krebs postulated that the metabolic sequence was not linear but cyclic and for this and other work in the field of intermediary metabolism Krebs was awarded the Nobel Prize in 1952.

Later work with isotopes confirmed the original hypothesis and revealed further details of the pathway. It was also shown that all the reactions of the cycle take place in the mitochondria. The metabolic pathway is called the Krebs' cycle after its discoverer or the tricarboxylic acid cycle (TCA) from the first intermediates formed.

The production of acetyl-CoA from pyruvate

OVERALL REACTION

Krebs originally postulated that the three-carbon pyruvate combined with the four-carbon oxaloacetate to form a seven-carbon intermediate, but later work showed that pyruvate is decarboxylated and that acetyl-CoA is the true substrate for the cycle. This reaction is an example of *oxidative decarboxylation* and can be summarized:

$$CH_3COCOO^- + H^+ + NAD^+ + CoA—SH$$

$$\downarrow$$

$$CH_3CO-S-CoA + CO_2 + NADH + H^+$$

PYRUVATE DEHYDROGENASE

This reaction is catalysed by pyruvate dehydrogenase, a complex containing three enzymes: pyruvate dehydrogenase (PDH), dihydrolipoyl dehydrogenase (DDH) and dihydrolipoyl transacetylase (DT) and the involvement of these enzymes and their cofactors in the formation of acetyl-CoA from pyruvate is shown in Fig. 8.14.

One of the low-molecular-weight cofactors not met before is *lipoic acid*:

$$\begin{array}{cc} \text{SH} & \text{SH} \\ | & | \\ \text{lipoic acid} \quad CH_2—CH_2—CH—CH_2—CH_2—CH_2—CH_2—COOH \end{array}$$

This can exist in the reduced, acetylated or oxidized form, all of which play a part in the decarboxylation of pyruvate (Fig. 8.14):

$$\begin{array}{ccc} \text{HS} & \text{CH}_3\text{CO—S} & \text{S} \\ \diagdown & \diagdown & |\diagdown \\ \text{lipoic acid} \quad \text{L} & \text{L} & \text{L} \\ \diagup & \diagup & |\diagup \\ \text{HS} & \text{HS} & \text{S} \\ \text{reduced} & \text{acetylated} & \text{oxidized} \end{array}$$

THE IMPORTANCE OF VITAMINS

Lipoic acid is one of the five coenzymes or prosthetic groups used by the pyruvate dehydrogenase complex but it is the only one that can be synthesized *de novo* by man. All of the other cofactors are synthesized from vitamins so the decarboxylation of

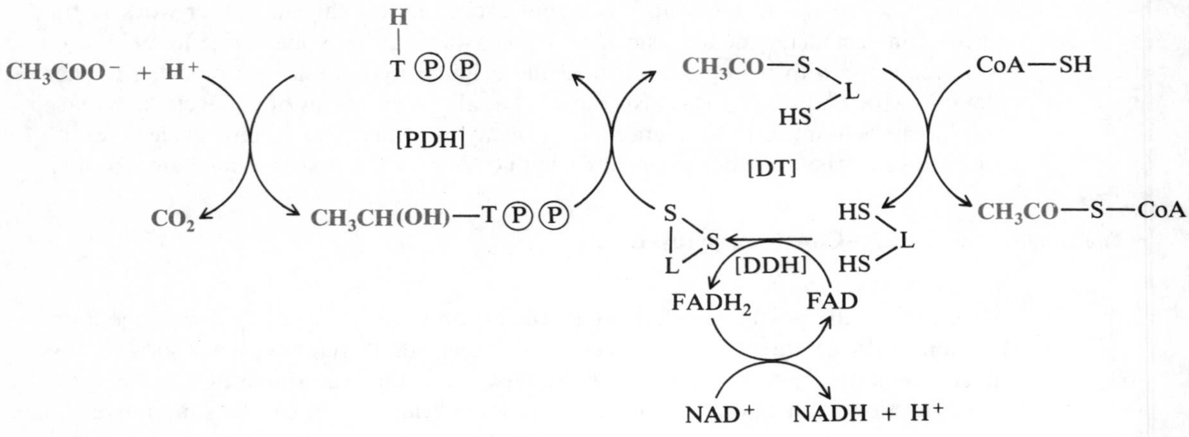

Figure 8.14 The formation of acetyl coenzyme A from pyruvate.

Enzymes

[PDH] = pyruvate dehydrogenase
[DT] = dihydrolipoyl transacylase
[DDH] = dihydrolipoyl dehydrogenase

Substrates and cofactors

T(P)(P) = thiamine pyrophosphate
FAD = flavin adenine dinucleotide
NAD^+ = nicotinamide adenine dinucleotide
CoA—SH = coenzyme A
L⟨SH / SH = lipoic acid (reduced form)

pyruvate depends very much on an adequate diet containing these vitamins (Table 8.2). The importance of these cofactors can be seen in the case of *beriberi*, a vitamin deficiency disease that develops following a lack of thiamine. People suffering from this disease have a raised blood pyruvate due to their reduced ability to convert pyruvate to acetyl-CoA. This particularly affects the nervous system since the brain can only use glucose as a metabolic fuel so that all of the acetyl-CoA has to come from pyruvate.

Table 8.2 The cofactors needed for oxidative decarboxylation and their vitamin precursors

Coenzyme or prosthetic group	Vitamin precursor
Flavin adenine dinucleotide	Riboflavin (B_2)
Thiamine pyrophosphate	Thiamine (B_1)
Nicotinamide adenine dinucleotide	Nicotinic acid
Coenzyme A	Pantothenic acid
Lipoic acid	—

Reactions of the Krebs' cycle

ENTRY OF ACETATE ($2c + 4c \rightarrow 6c$)

The first reaction of the Krebs' cycle is the formation of citrate from acetyl-CoA and oxaloacetate. This is catalysed by *citrate synthase* (CS) and is irreversible due to the hydrolysis of acetyl-CoA:

$$
\begin{array}{ccccccccc}
\text{CoA} & & & & & \text{COO}^- & & & \\
| & & & & & | & & & \\
\text{S} & & \text{O}=\text{C}-\text{COO}^- & \xrightarrow[\text{[CS]}]{\text{H}_2\text{O}} & \text{HO}-\text{C}-\text{COO}^- & + & \text{CoA} & + & \text{H}^+ \\
| & + & | & & | & & & \\
\text{C}=\text{O} & & \text{CH}_2 & & \text{CH}_2 & & \text{S} & \\
| & & | & & | & & | & \\
\text{CH}_3 & & \text{COO}^- & & \text{COO}^- & & \text{H} & \\
\end{array}
$$

acetyl – CoA   oxaloacetate   citrate   CoA

The citrate is then converted to isocitrate by the enzyme *aconitase* [A]:

citrate   *cis*-aconitate   isocitrate

OXIDATIVE DECARBOXYLATIONS

These irreversible reactions are similar to that seen for the decarboxylation of pyruvate in that decarboxylation is accompanied by the loss of carbon dioxide.

The first of these reactions is the conversion of isocitrate to 2-oxoglutarate ($6C \rightarrow 5C + 1C + [2H]$) which is catalysed by the enzyme *isocitrate dehydrogenase* (ICDH):

isocitrate   oxalosuccinate   2-oxoglutarate

This is followed by another decarboxylation when 2-oxoglutarate dehydrogenase [2-OGDH], a multienzyme complex similar to pyruvate dehydrogenase, catalyses the conversion of 2-oxoglutarate to succinyl-CoA ($5C \rightarrow 4C + 1C + [2H]$):

$$
\begin{array}{c}
COO^- \\
| \\
CH_2 \\
| \\
CH_2 \\
| \\
C=O \\
| \\
COO^-
\end{array}
+ CoA{-}SH
\quad\xrightarrow[\text{[2-OGDH]}]{NAD^+ \quad NADH+H^+}\quad
\begin{array}{c}
COO^- \\
| \\
CH_2 \\
| \\
CH_2 \\
| \\
C=O \\
| \\
S \\
| \\
CoA
\end{array}
+ CO_2
$$

2-oxoglutarate   succinyl − CoA

The hydrolysis of succinyl-CoA has a large negative ΔG and some of this energy is captured when GTP is formed from GDP + P_i. This reaction, which involves *substrate level phosphorylation*, is catalysed by succinyl-CoA synthetase [SCS] also known as succinate thiokinase:

$$
\begin{array}{c}
COO^- \\
| \\
CH_2 \\
| \\
CH_2 \\
| \\
C=O \\
| \\
S \\
| \\
CoA
\end{array}
\text{succinyl − CoA}
+ H_2O
\quad\xrightarrow[\text{[SCS]}]{P_i+GDP \quad GTP}\quad
\begin{array}{c}
COO^- \\
| \\
CH_2 \\
| \\
CH_2 \\
| \\
COO^-
\end{array}
+
\begin{array}{c}
H \\
| \\
S \\
| \\
CoA
\end{array}
+ H^+
$$

succinate   CoA

REGENERATION OF OXALOACETATE

The last phase of the Krebs' cycle is the regeneration of oxaloacetate and involves three enzymes—succinate dehydrogenase (SDH), fumarase (F) and malate dehydrogenase (MDH)—which catalyse two oxidative steps and a hydration:

$$
\begin{array}{c}
COO^- \\
| \\
H{-}C{-}H \\
| \\
H{-}C{-}H \\
| \\
COO^-
\end{array}
\xrightarrow[\text{[SCS]}]{FAD \quad FADH_2}
\begin{array}{c}
COO^- \\
| \\
C{-}H \\
\| \\
H{-}C \\
| \\
COO^-
\end{array}
\xrightarrow[\text{[F]}]{\pm H_2O}
\begin{array}{c}
COO^- \\
| \\
HO{-}C{-}H \\
| \\
H{-}C{-}H \\
| \\
COO^-
\end{array}
\xrightarrow[\text{[MDH]}]{\quad H^+ \atop NAP^+ \quad NADH}
\begin{array}{c}
COO^- \\
| \\
C=O \\
| \\
CH_2 \\
| \\
COO^-
\end{array}
$$

succinate   fumarate   malate   oxaloacetate

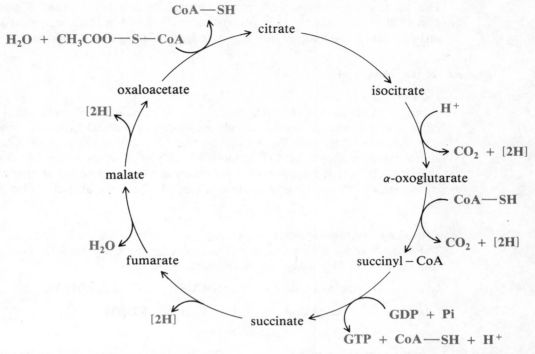

Figure 8.15 The essential features of the Krebs' cycle. The components of the cycle are shown in black and metabolites entering and leaving the cycle in red. The net reaction of the cycle is the oxidation of acetyl coenzyme A (Fig. 8.16).

The complete sequence of reactions is given in Fig. 8.15 which shows the cyclic nature of the metabolic pathway and the points where carbon dioxide and reducing equivalents in the form of $NADH + H^+$ and $FADH_2$ are produced.

REGULATION

The Krebs' cycle is similar to other metabolic pathways in that the rate depends on the first step and therefore the availability of oxaloacetate. Other metabolites also affect the rate at which the cycle functions and of these the ratios of NADH/NAD and ATP/ADP play an important part. Increased concentrations of NADH and ATP mean that the cell has enough energy to function and NADH inhibits pyruvate dehydrogenase, isocitrate dehydrogenase and 2-oxoglutarate dehydrogenase in mammalian systems and citrate synthase in Gram-negative bacteria. ATP inhibits pyruvate dehydrogenase and isocitrate dehydrogenase so that the cycle slows down. Conversely, when the NADH and ATP are low, the cycle speeds up.

Cellular metabolism is highly integrated and an example of this is the regulation by citrate of the supply of pyruvate to the Krebs' cycle. If there is an adequate supply of ATP, then isocitrate dehydrogenase is inhibited and this causes a build-up of citrate.

This intermediate then inhibits phosphofructokinase, a key enzyme of glycolysis, which reduces the supply of pyruvate to the cycle. This is a useful means for sparing carbohydrate when another substrate (e.g. fatty acids) is being oxidized.

Importance of the Krebs' cycle

A SOURCE OF ENERGY

During the operation of the cycle, there is no net gain or loss of oxaloacetate or the other intermediates so the equation for the oxidation of acetyl-CoA and the yield of ATP can be written as in Fig. 8.16. If pyruvate is taken as the starting point, then another three molecules of ATP are formed, making 15 for each turn of the cycle.

The Krebs' cycle, together with the electron transport chain, is therefore an important source of ATP and occupies a central role in the oxidation of metabolic fuels.

A SOURCE OF METABOLIC INTERMEDIATES

The Krebs' cycle is also a source of precursors for the biosynthesis of a number of compounds that play a key role in cellular metabolism:

$$oxaloacetate \longrightarrow GLUCOSE \text{ and } PYRIMIDINES$$

$$acetyl\text{-}CoA \longrightarrow FATS \text{ and } STEROIDS$$

$$succinyl\text{-}CoA \longrightarrow PORPHYRINS$$

A number of intermediates can also be converted to amino acids and this is dealt with in Section 8.6. This means that the Krebs' cycle is an *amphibolic pathway* (Gk *amphi* = both) because it serves both a catabolic and an anabolic function. The removal of

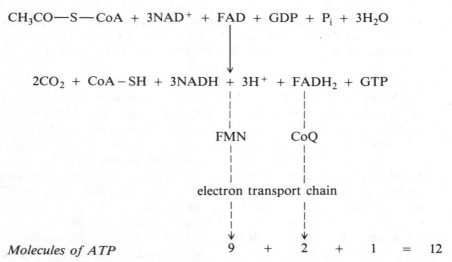

Figure 8.16   The oxidation of acetyl-CoA in the Krebs' cycle.

intermediates would have the effect of slowing down the cycle were it not for other reactions which replenish the metabolites. These are *anaplerotic pathways*, which means 'filling up': examples of these pathways are given in Section 9.1 on gluconeogenesis where oxaloacetate is synthesized from glycolytic intermediates.

SUMMARY OF THE CYCLE

The cycle can be summarized as follows:

Balance of cycle: $CH_3COSCoA + 3H_2O \longrightarrow 2CO_2 + 8[H] + CoASH$

Electron transport: $\qquad\qquad 8[H] + 2O_2 \longrightarrow 4H_2O$

Net reaction: $\qquad CH_3COSCoA + 2O_2 \longrightarrow 2CO_2 + H_2O + CoASH$

These equations show the essential features of the Krebs' cycle.

Catalyst   The cycle is essentially a catalyst for the terminal oxidation of acetyl residues.

Source of oxygen   The cycle uses water as a source of oxygen for the formation of carbon dioxide and not oxygen. Molecular oxygen only comes in when the reduced coenzymes are reoxidized in the electron transport chain.

8.6   Amino acids: a minor metabolic fuel

Glucose and fatty acids are the major metabolic fuels used by organisms and the oxidation of amino acids normally provides only a small part of their energy requirements. The main function of the amino acids as discussed earlier is to provide the building blocks for the construction of proteins but the turnover of proteins in the cell means that some amino acids are always available for oxidation. Sometimes the amino acids make a greater contribution than usual to the energy requirements of the body, for example when there is an excessive intake of protein in the diet or when the normal metabolic fuels are unavailable due to starvation or disease. Recent studies have shown that in some mammalian organs the oxidation of particular amino acids is an important source of energy. For example, the small intestine oxidizes the acidic amino acids, aspartate and glutamate and their corresponding amides, asparagine and glutamine and during starvation glutamine from the blood can provide as much as 30 per cent of the energy needed by the intestinal cells. Another example is that of muscle which plays an active part in the catabolism of branched chain amino acids such as valine.

Amino acids from dietary protein

HYDROLYSIS OF PROTEIN

Protein in the diet is broken down to peptides then amino acids in the gastrointestinal tract by a range of proteolytic enzymes. Digestion commences in the stomach where the acid environment (pH 2) denatures the proteins and renders them more susceptible to enzymic action. Peptide bonds in the interior of the molecule are first attacked by *endopeptidases*. These are secreted as inactive *pro-enzymes* or *zymogens* and are activated by the removal of a small peptide fragment. Such a procedure ensures that the cells producing the enzyme are protected and that dietary and not cellular proteins are broken down by the endopeptidases.

Pro-enzyme	*Activator*	*ENDOPEPTIDASE*
Pepsinogen	$\xrightarrow[\text{pepsin}]{\text{H}^+}$	PEPSIN
Trypsinogen	$\xrightarrow[\text{trypsin}]{\text{enteropeptidase}}$	TRYPSIN
Chymotrypsinogen	$\xrightarrow[\text{trypsin}]{}$	CHYMOTRYPSIN
Pro-elastase	$\xrightarrow[\text{trypsin}]{}$	ELASTASE

These enzymes attack specific peptide bonds (Table 8.3), so the effect of the combined action of pepsin in the stomach and the pancreatic enzymes in the duodenum is to break down the dietary proteins into smaller peptide fragments. Digestion is then completed in the small intestine by *exopeptidases* which hydrolyse the peptide bonds

Table 8.3   The specificity of the endopeptidases for peptide bonds
[----A—CO—NH—B-----]

Proteolytic enzyme	Source	Amino acids affected	
		A	B
Pepsin	Gastric mucosa	Phe Try Tyr Met Leu	Any
Trypsin	Pancreas	Basic Arg Lys	Any
Chymotrypsin	Pancreas	Aromatic Phe Tyr Try	Any except Asp and Glu
Elastase	Pancreas	Neutral aliphatic Gly Val Leu Ileu	Any

sequentially from the carboxyl end (carboxypeptidases) or amino end (aminopeptidases) of the polypeptide chain to give amino acids.

THE ABSORPTION OF AMINO ACIDS

The absorption of amino acids in the small intestine is similar to that of glucose. The amino acid and Na^+ are bound by a specific carrier in the brush border, transported across the membrane and released into the mucosal cell. Amino acids enter cells against a concentration gradient and the energy for this active transport is provided by the entry of Na^+ into the cells down its concentration gradient. There are three transport systems for neutral, basic and acidic amino acids and a fourth for glycine and the imino acids.

There is also another transport system in which the amino acid binds to glutathione, is carried across the membrane by the enzyme γ-glutamyltransferase and the glutathione regenerated in the *γ-glutamyl cycle*. However, the importance of this mechanism is still being debated since it involves the hydrolysis of three molecules of ATP and is therefore energetically costly.

Finally, some amino acids are absorbed from the gut as part of small peptides which are then hydrolysed by peptidases in the brush border.

THE AMINO ACID POOL

The amino acids pass into the capillaries and are transported in the blood to the tissue where they contribute to the cellular pool. Amino acids cannot be stored to any great extent and essential amino acids cannot be synthesized by the organism, so the

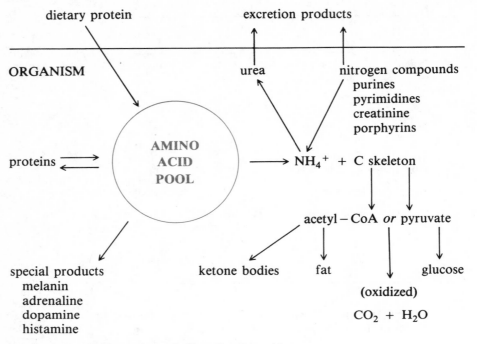

ENVIRONMENT

dietary protein   excretion products

ORGANISM   urea   nitrogen compounds
                                                      purines
                                                      pyrimidines
                                                      creatinine
                                                      porphyrins

proteins ⇌   **AMINO
                      ACID
                      POOL**   → NH_4^+ + C skeleton

                                      acetyl – CoA *or* pyruvate

special products   ketone bodies   fat   glucose
  melanin
  adrenaline   (oxidized)
  dopamine
  histamine CO_2 + H_2O

Figure 8.17 The general metabolism of amino acids.

synthesis of new protein is dependent upon an adequate supply of protein in the diet. There is a rapid turnover of amino acids in the pool and the metabolic fate of these compounds is shown in Fig. 8.17.

The removal of nitrogen

Before amino acids can be used as a metabolic fuel, the nitrogen is first removed and there are three main ways that this occurs:

1. D-AMINO ACID OXIDASE

D-Amino acid oxidase (D-AAO) was the first enzyme to be discovered that catalysed the removal of nitrogen from amino acids. The oxidase is a flavoprotein enzyme which acts on D amino acids to form the corresponding keto acid, hydrogen peroxide and ammonia:

$$_3\overset{+}{H}N-\underset{|}{\overset{R}{\underset{COO^-}{\overset{|}{C}}}}-H + H_2O + O_2 \xrightarrow{[D-AAO]} \underset{|}{\overset{R}{\underset{COO^-}{\overset{|}{C}}}}=O + H_2O_2 + NH_4^+$$

D amino acid   keto acid

Hydrogen peroxide is highly toxic but this is immediately broken down to water and oxygen by catalase which lies close to the D amino acid oxidase in peroxisomes.

The oxidase is present in bacteria where there is a high turnover of D amino acids and it is also found in kidney and liver where it deaminates D amino acids that have passed into the bloodstream from the breakdown of bacteria in the colon. This is probably a defence mechanism to remove D amino acids which might otherwise interfere with the metabolism of the L amino acids that are normally used by animals.

Animal tissues also contain an L amino acid oxidase but the activity is so low that it plays only a minor role in the deamination of the L amino acids.

2. TRANSAMINATION

Removal of the nitrogen from L amino acids in animals takes place by transamination in which an aminotransferase (AT) catalyses the reaction of an amino acid with 2-oxoglutarate to form the corresponding oxo acid and L-glutamate:

$$
\begin{array}{cccc}
\text{R} & \text{COO}^- & & \text{R} & \text{COO}^- \\
| & | & [AT] & | & | \\
\text{H—C—}\overset{+}{\text{NH}}_3 \;+\; & \text{CH}_2 & \rightleftharpoons & \text{C}=\text{O} & \text{CH}_2 \\
| & | & & | & | \\
\text{COO}^- & \text{CH}_2 & & \text{COO}^- & \text{CH}_2 \\
& | & & & | \\
& \text{C}=\text{O} & & & \text{H—C—}\overset{+}{\text{NH}}_3 \\
& | & & & | \\
& \text{COO}^- & & & \text{COO}^-
\end{array}
$$

L amino acid   2-oxoglutarate   keto acid   L-glutamate

All aminotransferases contain a tightly bound prosthetic group which is derived from *vitamin B₆* and acts as a carrier of the amino group. This is present as *pyridoxal phosphate* or *pyridoxamine phosphate* and during transamination there is a reversible interconversion of the two forms:

pyridoxal phosphate   pyridoxamine phosphate

Transamination in mammals takes place with all amino acids except lysine and threonine although the activities shown by serine, methionine, histidine and phenylalanine *in vivo* are quite low.

3. GLUTAMATE DEHYDROGENASE (GDH)

This mitochondrial enzyme catalyses the removal of nitrogen from L-glutamate as NH_4^+ to give 2-oxoglutarate and the reduction of NAD^+ or $NADP^+$ to NADH or NADPH. The reaction catalysed by GDH is a good example of oxidative deamination:

$$\text{L-glutamate} + NAD^+ + H_2O \underset{\text{[GDH]}}{\rightleftharpoons} \text{2-oxoglutarate} + NADH + NH_4^+$$

Glutamate dehydrogenase is freely reversible and the combination of this enzyme with an aminotransferase means that nearly all of the L amino acids can be deaminated. The flow of nitrogen is shown in bold in the following metabolic sequence:

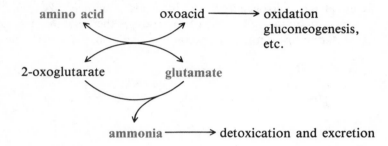

The carbon skeleton as a metabolic fuel

The ammonia released by deamination is highly toxic and is rapidly removed by metabolism. This detoxification of ammonia is extremely important and how this takes place is considered later in the book. For the moment we shall concentrate on the metabolic fate of the carbon skeleton that remains after deamination, since it is this part of the molecule that can act as a metabolic fuel.

GLUCOGENIC AMINO ACIDS

If amino acids are fed to a starving or diabetic animal there is an increase in either the blood glucose or ketone bodies or sometimes both. Most amino acids cause an increase in blood glucose and these are called *glucogenic amino acids* while a few give rise to ketone bodies and these are known as *ketogenic amino acids*. In the case of the glucogenic amino acids, the carbon skeleton is converted to pyruvate or an intermediate of the Krebs' cycle and then to glucose or glycogen. This is known as *gluconeogenesis* and the details of this important process are discussed in Section 9.1.

Alanine is a good example of a glucogenic amino acid and this is converted to pyruvate by alanine aminotransferase [AlaAT]:

$$\text{L-alanine} + \text{2-oxoglutarate} \underset{\text{[AlaAT]}}{\rightleftharpoons} \text{pyruvate} + \text{L-glutamate}$$

Aspartate is similar and forms oxaloacetate under the influence of aspartate aminotransferase [AspAT]:

$$\text{L-aspartate} + \text{2-oxoglutarate} \xrightleftharpoons{\text{[AspAT]}} \text{oxaloacetate} + \text{L-glutamate}$$

Most of the glucogenic amino acids are non-essential and can be synthesized by a reversal of the reactions involved in their deamination. However, some are essential and cannot be synthesized by the organism due to an irreversible step in their catabolism.

KETOGENIC AMINO ACIDS

These amino acids are metabolized to acetyl-CoA and this involves an irreversible step. They cannot therefore be synthesized by the organism and must be supplied in the diet. Leucine and lysine are purely ketogenic while phenylalanine, tyrosine, tryptophan and isoleucine are both ketogenic and glucogenic.

THE ROLE OF THE KREBS' CYCLE

The points at which the carbon skeletons are fed into the Krebs' cycle are shown in Fig. 8.18. Acetyl-CoA from the ketogenic amino acids can be oxidized in the cycle but

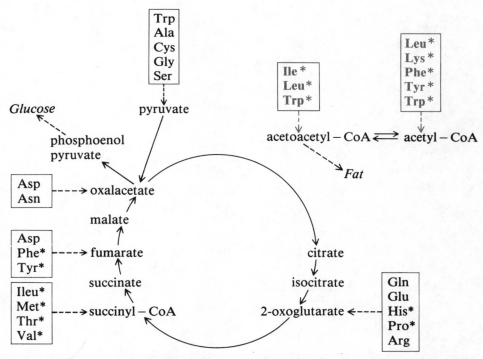

Figure 8.18 Amino acids and the Krebs' cycle. The glucogenic amino acids are shown in black and the ketogenic amino acids in red; essential amino acids are marked *.

this cannot happen directly with the glucogenic amino acids. Krebs'-cycle intermediates must first leave the cycle and then enter as acetyl-CoA before they can be oxidized. The mechanism for this is discussed in the Section 9.1 on glucose synthesis.

Finally, it is worth noting that the Krebs' cycle, as shown in Fig. 8.18, is the meeting point for the metabolism of three major classes of compound, namely, carbohydrate, fat and protein. The function of the cycle is therefore the terminal oxidation of anything that can give rise to acetyl-CoA.

9. Biosynthesis

9.1 The synthesis of glucose: gluconeogenesis

The importance of glucose synthesis in animals

Glucose is an important metabolic fuel and in a well-fed animal, an adequate supply is obtained from the breakdown of carbohydrates in the diet. During periods of malnourishment or fasting, the blood glucose is maintained by the breakdown of liver glycogen but these stores soon become depleted. Under these circumstances, fat is mobilized from adipose tissue and fatty acids become the dominant metabolic fuel. However, some tissues such as the brain and human erythrocytes are dependent upon glucose.

THE BRAIN

The human brain, although it accounts for only 2 per cent of the total body weight, consumes 70 per cent or more of the total glucose used by the body and under normal circumstances the brain is unable to use any other metabolic fuel. During periods of prolonged and severe starvation, the brain begins to use 3-hydroxybutyrate as a fuel and after six weeks without food, the oxidation of this ketone body is able to supply as much as 70 per cent of the energy requirements of the brain with the other 30 per cent coming from the oxidation of glucose.

The need for glucose can be dramatically seen in the effect of a fall in the blood sugar which if severe leads to mental confusion, hallucinations, coma and eventually death. Glucose therefore must continue to be supplied to the brain even during starvation.

HUMAN ERYTHROCYTES

Mature erythrocytes also can only use glucose as a metabolic fuel with 90 per cent going to lactate via glycolysis and 10 per cent being metabolized via the pentose phosphate pathway. Glucose metabolism gives rise to a number of important intermediates which play an important part in maintaining the structural integrity and normal physiological function of these cells and some examples are given below:

2,3-Bisphosphoglycerate	Controls the oxygen dissociation curve of haemoglobin.
ATP	Drives the Na^+ pump.
NADH and NADPH	Keeps glutathione reduced (GSH).
Glutathione (GSH)	Protects haemoglobin and the cell membrane from oxidation.

RUMINANTS

Cows and other ruminants are able to digest grass in the first compartment of their stomach called the *rumen*. Here bacteria hydrolyse the cellulose and convert nearly all the glucose produced to acetate, propionate and butyrate. These short-chain fatty acids can be used as metabolic fuels but cows, like other animals, need to maintain a reasonable level of blood glucose for catabolism and also for the synthesis of lactose, the sugar present in milk. An adequate supply of glucose is therefore particularly important in ruminants during pregnancy and lactation.

GLUCONEOGENESIS

These examples clearly show the importance of glucose in particular cases but some carbohydrate metabolism needs to take place to avoid ketosis even in those cells that can oxidize fatty acids. Glucose therefore must be provided and during periods of starvation this is synthesized from non-carbohydrate sources by *gluconeogenesis*.

The irreversible steps of glycolysis

The synthesis of glucose from pyruvate or lactate is essentially the reverse of glucose breakdown and most of the enzymes involved are the same as those in glycolysis (Figs. 8.5 and 8.6). However, there are some steps in the glycolytic pathway which are irreversible and different enzymes are used in gluconeogenesis to reverse these changes.

PHOSPHOENOLPYRUVATE (PEP) TO PYRUVATE

The first step needing a by-pass is the interconversion of PEP and pyruvate. In glycolysis the formation of pyruvate from PEP is catalysed by pyruvate kinase (PK).

$$\text{phosphoenolpyruvate} + \text{ADP} \xrightarrow[\text{Mg}^{2+}]{\text{[PK]}} \text{pyruvate} + \text{ATP}$$

This reaction has a high negative free energy change ($\Delta G = -26\,\text{kJ mol}^{-1}$) and is

lysine residue   carboxybiotin

therefore irreversible so that energy needs to be put into the system for the phosphorylation of pyruvate to take place. The driving force for the reverse process is the carboxylation–decarboxylation of pyruvate and the hydrolysis of ATP and GTP. The first step is the carboxylation of pyruvate to form oxaloacetate, a reaction which is catalysed by the enzyme pyruvate carboxylase (PC). The carbon dioxide is present as HCO_3^- and this is 'activated' by the vitamin biotin which is bound to a lysine residue on the enzyme.

The second stage is the decarboxylation of oxaloacetate coupled with its phosphorylation by GTP; this is catalysed by the enzyme phosphoenolpyruvate carboxykinase (PEP-CK). The net result of these two reactions is that pyruvate is phosphorylated via oxaloacetate to form PEP. The hydrolysis of ATP and GTP means that the overall free energy change is negative ($G = -22 \, \text{kJ mol}^{-1}$) so the reaction proceeds spontaneously:

$$\text{pyruvate} + CO_2 + \text{ATP} \xrightarrow[Mg^{2+}]{[PC]} \text{oxaloacetate} + \text{ADP} + P_i$$

$$\text{oxaloacetate} + \text{GTP} \xrightarrow[Mg^{2+}, Mn^{2+}]{[PEP-CK]} \text{phosphoenolpyruvate} + CO_2 + \text{GDP}$$

$$\text{pyruvate} + \text{ATP} + \text{GTP} \longrightarrow \text{phosphoenolpyruvate} + \text{ADP} + \text{GDP} + P_i$$

FRUCTOSE-6-PHOSPHATE TO FRUCTOSE-1,6-BISPHOSPHATE

The second irreversible step of glycolysis with a high negative G ($-24 \, \text{kJ mol}^{-1}$) is the phosphorylation of fructose-6-phosphate to form fructose-1,6-bisphosphate. The reaction is catalysed by phosphofructokinase (PFK) and driven by the hydrolysis of ATP:

$$\text{fructose-6-phosphate} + \text{ATP} \xrightarrow[Mg^{2+}]{[PFK]} \text{fructose-1,6-bisphosphate} + \text{ADP}$$

The reverse reaction in gluconeogenesis is brought about by fructose-1,6-bisphosphatase (FBP) with a ΔG of $-8.5 \, \text{kJ mol}^{-1}$:

$$\text{fructose-1,6-bisphosphate} + H_2O \xrightarrow{[FBP]} \text{fructose-6-phosphate} + P_i$$

GLUCOSE TO GLUCOSE-6-PHOSPHATE

The third step in glycolysis which needs to be by-passed is the conversion of glucose to glucose-6-phosphate. The phosphorylation of glucose by hexokinase (HK) and glucokinase (GK) is irreversible due to the large free energy change arising from the hydrolysis of ATP:

$$\text{glucose} + \text{ATP} \xrightarrow[Mg^{2+}]{[HK][GK]} \text{glucose-6-phosphate} + \text{ADP}$$

The reverse reaction in gluconeogenesis is therefore catalysed by a different enzyme, glucose-6-phosphatase (G-6-Pase) with a favourable ΔG ($-5\,kJ\,mol^{-1}$):

$$\text{glucose-6-phosphate} + H_2O \xrightarrow{\text{[G-6-pase]}} \text{glucose} + P_i$$

Energy changes in glycolysis and gluconeogenesis

Gluconeogenesis is energetically costly and the synthesis of one molecule of glucose from two molecules of pyruvate consumes four molecules of ATP and two of GTP (the equivalent of six ATP):

$$2\ \text{pyruvate} + 4ATP + 2GTP + 2NADH + 2H^+ + 6H_2O$$
$$\downarrow \text{(gluconeogenesis)}$$
$$\text{glucose} + 2NAD^+ + 4ADP + 2GDP + 6P_i$$

This compares with only two ATP generated from glycolysis:

$$\text{glucose} + 2ADP + P_1 + 2NAD^+$$
$$\downarrow \text{(glycolysis)}$$
$$2\ \text{pyruvate} + 2ATP + 2NADH + 2H^+ + 2H_2O$$

Non-carbohydrate sources of glucose

LACTATE

Lactate produced by erythrocytes and rapidly contracting muscles is an important substrate for gluconeogenesis. During the period immediately following a sprint or strenuous exercise, lactate diffuses from the muscles and is carried by the blood to the liver where it is acted on by lactate dehydrogenase (LDH):

$$CH_3CH(OH)COO^- + NAD^+ \underset{}{\overset{\text{[LDH]}}{\rightleftarrows}} CH_3COCOO^- + NADH + H^+$$

The high concentration of lactate and the high ratio of NAD^+ to NADH in the liver ensures that the lactate is readily converted to pyruvate and to glucose. This means that glycogen in muscle can contribute to the blood sugar albeit in a roundabout fashion. If the glucose returns to the muscle then glycolysis in muscle is connected to gluconeogenesis in liver in a metabolic sequence known as the Cori cycle (Fig. 9.1). The breakdown of glucose to lactate generates protons and the conversion of lactate to glucose is important in protecting the body against the adverse effects of lactic acidosis:

$$C_6H_{12}O_6 \underset{\text{gluconeogenesis}}{\overset{\text{glycolysis}}{\rightleftarrows}} 2CH_3CH(OH)COO^- + 2H^+$$

GLYCEROL

Glycerol from the hydrolysis of triacylglycerols can be phosphorylated and converted to dihydroxyacetone phosphate, one of the intermediates of glycolysis. From here it

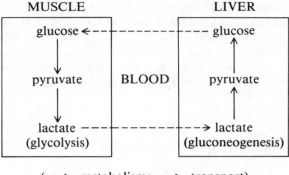

MUSCLE   LIVER

glucose ←--------------------- glucose

pyruvate   BLOOD   pyruvate

lactate -------------------→ lactate
(glycolysis)   (gluconeogenesis)

(⟶ metabolism; --→ transport)

Figure 9.1 The Cori cycle. A metabolic cycle showing the relationship between lactate and glucose in muscle, blood and liver.

can be oxidized to pyruvate or converted to glucose in gluconeogenesis and details of these reactions are given in Fig. 8.8.

AMINO ACIDS

Some glucogenic amino acids are metabolized to pyruvate while others are converted to intermediates of the Krebs' cycle (Fig. 8.18). Those amino acids that form pyruvate directly can be oxidized in the cycle or converted to glucose by gluconeogenesis. The carbon skeletons of those amino acids that give rise to the Krebs'-cycle intermediates have first to leave the cycle before they can be metabolized further. The usual exit is from oxaloacetate which can be converted to phosphoenolpyruvate by PEP-carboxykinase as discussed earlier. Yet another exit is through malate which can be oxidized to pyruvate by the *malic enzyme* (ME) but the evidence suggests that the main function of this enzyme is to provide NADPH for the synthesis of fatty acids since it is induced under lipogenic conditions:

$$
\begin{array}{l}
COO^- \\
| \\
CH_2 \\
| \qquad\qquad\qquad\qquad [ME] \\
CH(OH) + NADP^+ \rightleftharpoons \\
| \\
COO^-
\end{array}
\qquad
\begin{array}{l}
CH_3 \\
| \\
CO + NADPH + H^+ + CO_2 \\
| \\
COO^-
\end{array}
$$

The glyoxylate cycle in plants

CARBOHYDRATE FROM FAT?

Animals are unable to synthesize glucose from acetyl-CoA, the breakdown product of fatty acids, because the conversion of pyruvate to acetyl-CoA is irreversible. At first sight another possible route for glucose synthesis from acetyl-CoA is via the intermediates of the Krebs' cycle since these can and do give rise to glucose. However, when acetyl-CoA enters the cycle, two carbons are liberated as carbon dioxide in the first two steps of oxidative decarboxylation so there is no net synthesis of the four-carbon intermediates which follow and therefore no net synthesis of glucose.

Some plants can synthesize four-carbon and six-carbon compounds from two-carbon units and they do this by cutting out the two decarboxylation steps of the Krebs' cycle. Two enzymes are involved in this: isocitrate lyase [IL] and malate synthase [MS]:

1.

$$
\begin{array}{l}
CH_2COO^- \\
| \\
H-C-COO^- \\
| \\
HO-C-H \\
| \\
COO^-
\end{array}
\quad
\underset{}{\overset{[IL]}{\rightleftharpoons}}
\quad
\begin{array}{l}
CH_2COO^- \\
| \\
{}^-OOCCH_2
\end{array}
\quad + \quad
\begin{array}{l}
COO^- \\
| \\
CHO
\end{array}
$$

  isocitrate   succinate   glyoxylate

2.

$$
\begin{array}{l}
COO^- \\
| \\
CHO
\end{array}
\quad + \quad CH_3CO-S-CoA + H_2O
\quad
\underset{}{\overset{[MS]}{\rightleftharpoons}}
\quad
\begin{array}{l}
COO^- \\
| \\
HO-C-H \\
| \\
H-C-H \\
| \\
COO^-
\end{array}
+ \; CoA-SH + H^+
$$

glyoxylate   acetyl-CoA   malate

Malate is metabolized to isocitrate by the enzymes of the Krebs' cycle so the net result is that two two-carbon units as acetate are converted to one four-carbon unit in the form of succinate. In this pathway therefore there is a net synthesis of Krebs'-cycle intermediates which can then be converted to hexose units which are used to synthesize plant carbohydrates.

THE FUNCTION OF THE GLYOXYLATE CYCLE

The *glyoxylate cycle* (Fig. 9.2) is particularly important during the germination of certain plants whose seedlings have food stores of fat rather than the usual carbohydrate. Examples of such plants include peanuts, sunflowers and castor beans. In the seeds of these plants, the acetyl-CoA units from the breakdown of fatty acids are used to synthesize carbohydrates needed for the early stages of growth before photosynthesis has started. In plants, the enzymes for the glyoxylate cycle are found in special organelles called *glyoxysomes* which are a specialized form of peroxisomes.

The cycle is also found in micro-organisms and explains how the simple forms of life such as bacteria, fungi and algae can grow on a simple two-carbon substrate such as acetate.

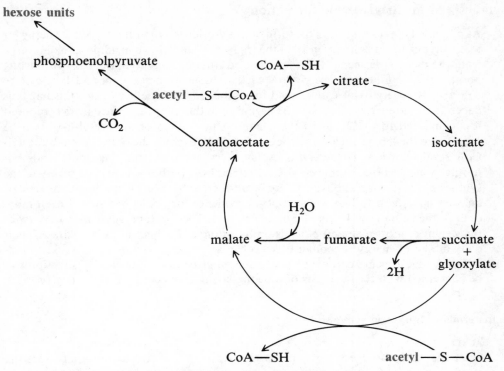

Figure 9.2 The essential features of the glyoxylate cycle.

9.2   Synthesis of fat: triglyceride formation

As discussed previously, fat in the form of triglyceride is a long-term store of energy for many organisms. In animals most of the triglyceride laid down in adipose tissue comes from the diet but fat can also be synthesized from non-fat compounds. The starting point for fat synthesis is acetyl-CoA so that basically any compound that can be converted to acetyl-CoA can be used to synthesize fat. Whether or not this happens depends on the energy balance of the organism. If the energy obtained from the food is low compared to the needs of the organism then most of the acetyl-CoA will be oxidized in the Krebs' cycle but if the energy intake exceeds the output then the acetyl-CoA is laid down as fat. Acetyl-CoA can be synthesized from the amino acids but normally most of the excess acetyl-CoA comes from carbohydrate. The reality of the synthesis of fat from carbohydrate can be attested to by most of us who at one time or another have tried to lose weight but the evidence for this conversion is even more clear in the case of the Japanese Sumo wrestlers. These rotund gentlemen have twice the weight of most men and put this on by eating a diet high in carbohydrate. Their daily intake of energy is enough to keep five average men going for a week!

Triacylglycerols are carboxylic acid esters of glycerol and most of this section will be concerned with the synthesis of the long-chain fatty acid part of the triglycerides.

The raw materials of fatty acid synthesis

ACETYL-COA

Since the start of this century it has been suspected that two-carbon units were the starting material for the synthesis of fatty acids but it was not until isotopes were widely available that this was confirmed and the biosynthetic pathways worked out. Fatty acid synthesis takes place in the cytoplasm but acetyl-CoA is generated in the mitochondria so the first step in the pathway is the need to transfer acetyl-CoA from the mitochondria to the cytoplasm. The mitochondrial inner membrane is impermeable to acetyl-CoA and transport occurs via the citrate–malate shuttle in the membrane and enzymes in these two compartments (Fig. 9.3).

NADPH

The other 'raw material' needed for fatty acid synthesis is a source of electrons. For just as reducing equivalents are released during the oxidation of fatty acids to acetyl-CoA so reducing power is needed for the reduction of acetyl-CoA to fatty acids. NADH cannot be used since the ratio of $NADH/NAD^+$ in the cytoplasm is about 1/1000 which strongly favours oxidation. However, NADPH is highly suitable since the ratio of $NADPH/NADP^+$ is about 100/1 which favours reduction.

NADPH and anabolism   Incidentally the difference in the ratio of the reduced and oxidized forms of the two coenzymes serves as a metabolic division of labour with NAD^+ mainly involved in catabolism and NADPH in anabolism.

Other examples of how NADPH is used in anabolic reactions include not only fatty

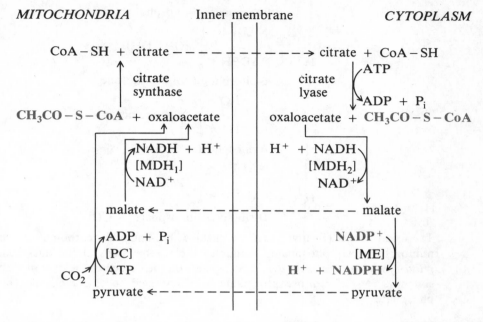

$MITOCHONDRIA$ Inner membrane $CYTOPLASM$

[MDH$_1$] and [MDH$_2$] = isoenzymes of malate dehydrogenase
[ME] = malic enzyme, [PC] = pyruvate carboxylase

($--\rightarrow$ transport; $\longrightarrow$ metabolism)

Figure 9.3 The transport of acetyl groups from the mitochondria to the cytoplasm and the generation of NADPH needed for the synthesis of fatty acid.

acid synthesis in adipose tissue and liver but the synthesis of steroids by the adrenal cortex, testes and ovaries and hydroxylation reactions in the adrenal medulla and nervous tissue.

The pentose phosphate pathway   Most of the NADPH comes from the pentose phosphate pathway of glucose metabolism which can be described in two phases.

I.   Oxidation of glucose-6-phosphate:

$$\text{glucose-6-phosphate} \longrightarrow \text{pentose-5-phosphate} + CO_2$$

II.   Regeneration of hexose phosphate:

$$6 \text{ pentose-5-phosphate} \longrightarrow 5 \text{ hexose-6-phosphate}$$

The first stage (I) generates two molecules of NADPH from the reactions catalysed by glucose-6-phosphate dehydrogenase (GP-DH) and 6-phosphogluconate dehydrogenase (PG-DH).

$$\beta\text{-D-glucose-6-phosphate}$$

NADP$^+$ ⟍
H$^+$ + NADPH ⟋ [GP – DH]

$$6\text{-phosphoglucono-}\delta\text{-lactone}$$

H$_2$O ⟍
H$^+$ ⟋ [gluconolactonase]

$$6\text{-phospho-D-gluconate}$$

NADP$^+$ ⟍
H$^+$ + NADPH ⟋ [PG – DH]

$$\text{D-ribulose-5-phosphate} + CO_2$$

The second phase (II) involves a complicated shuffling of the carbon atoms between a number of sugar phosphates similar to the reverse steps of the dark reaction of photosynthesis. The exact sequence depends on the tissue but the overall effect is that six molecules of pentose phosphate are converted to five molecules of hexose phosphate.

Synthesis of palmitate

THE FATTY ACID SYNTHASE SYSTEM

There are seven enzymes involved in the synthesis of fatty acids and they are present in the cytoplasm in the form of a multienzyme complex. In prokaryotes these enzymes can be separated and the reactions they catalyse studied individually but in eukaryotes all the reactions take place on one large multifunctional enzyme complex. A key part of this complex is the *acyl carrier protein* (ACP) which contains an —SH group at the end of a long arm of phosphopantetheine. This group is mobile and swings the growing fatty acid chain from one enzyme to the next. The acyl carrier protein contains a second —SH group from cysteine that is fixed and this also plays an important part in the synthesis.

THE LOADING OF THE ACYL GROUPS

The first step in this process is the synthesis of malonyl-CoA from acetyl-CoA catalysed by *acetyl-CoA carboxylase* (AcCoA-C). This, like other carboxylation reactions, has biotin as a prosthetic group and uses HCO_3^- as the source of carbon dioxide:

$$ATP + \begin{array}{c} CH_3 \\ | \\ C=O \\ | \\ S \\ | \\ CoA \end{array} + CO_2 \xrightarrow[\text{biotin}]{[AcCoA - C]} \begin{array}{c} COO^- \\ | \\ CH_2 \\ | \\ C=O \\ | \\ CoA \end{array} + H^+ + ADP + P_i$$

acetyl — CoA   malonyl — CoA

This is the control site in the process since malonyl-CoA can only be used for fatty acid synthesis.

The two —SH groups are then loaded: the first with an acetyl group from acetyl-CoA and the second with a malonyl group from malonyl-CoA (Fig. 9.4).

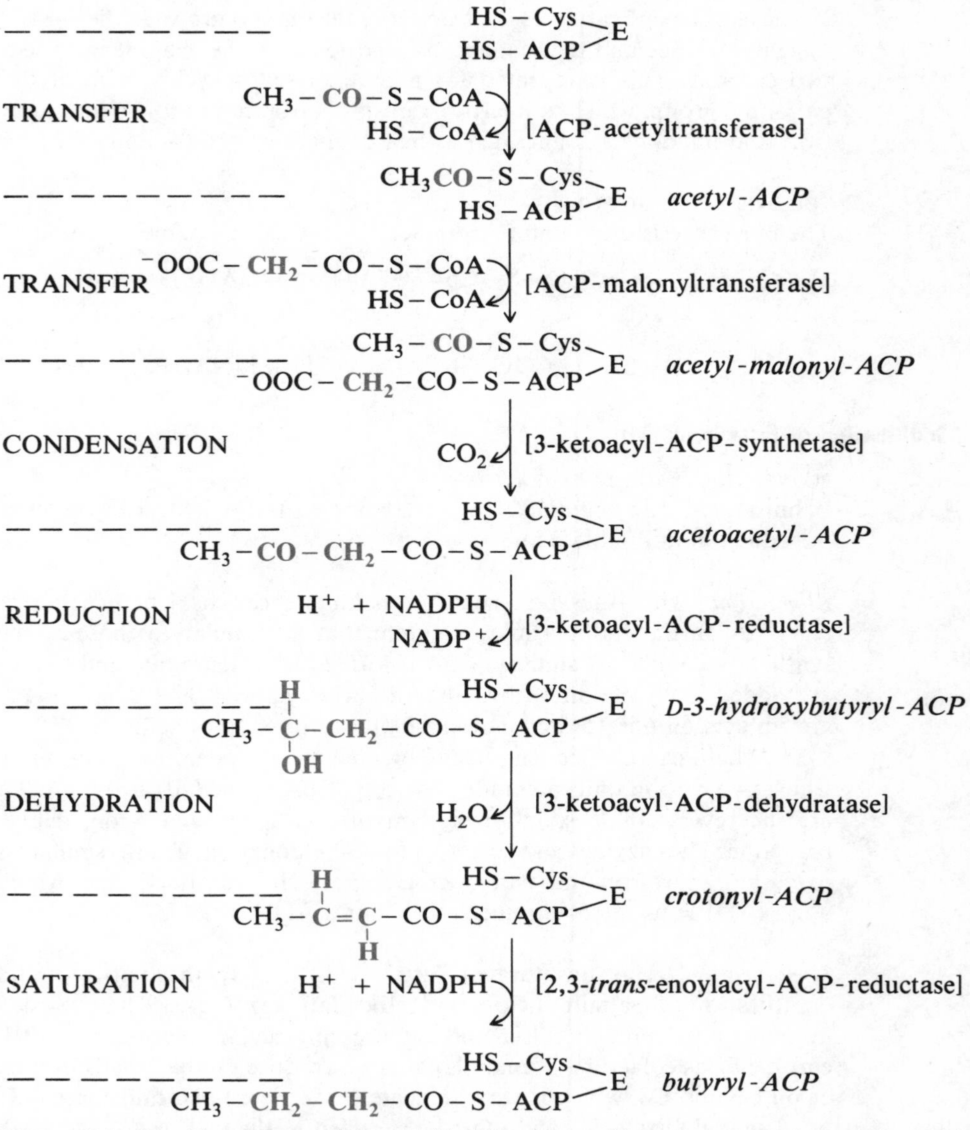

$$
\begin{array}{ll}
\text{— — — — — — — —} & \begin{array}{l} HS-Cys \\ HS-ACP \end{array} \!\!\! \Big\rangle E \\[2mm]
\text{TRANSFER} \quad \begin{array}{l} CH_3-CO-S-CoA \\ HS-CoA \end{array} \!\!\! \Big\rangle \quad [\text{ACP-acetyltransferase}] \\[2mm]
\text{— — — — — — — —} & \begin{array}{l} CH_3CO-S-Cys \\ HS-ACP \end{array} \!\!\! \Big\rangle E \quad \textit{acetyl-ACP} \\[2mm]
\text{TRANSFER} \quad \begin{array}{l} {}^-OOC-CH_2-CO-S-CoA \\ HS-CoA \end{array} \!\!\! \Big\rangle \quad [\text{ACP-malonyltransferase}] \\[2mm]
\text{— — — — — — —} & \begin{array}{l} CH_3-CO-S-Cys \\ {}^-OOC-CH_2-CO-S-ACP \end{array} \!\!\! \Big\rangle E \quad \textit{acetyl-malonyl-ACP} \\[2mm]
\text{CONDENSATION} \quad CO_2 \!\!\! \Big\rangle \quad [\text{3-ketoacyl-ACP-synthetase}] \\[2mm]
\text{— — — — — — —} & \begin{array}{l} HS-Cys \\ CH_3-CO-CH_2-CO-S-ACP \end{array} \!\!\! \Big\rangle E \quad \textit{acetoacetyl-ACP} \\[2mm]
\text{REDUCTION} \quad \begin{array}{l} H^+ + NADPH \\ NADP^+ \end{array} \!\!\! \Big\rangle \quad [\text{3-ketoacyl-ACP-reductase}] \\[2mm]
\end{array}
$$

DEHYDRATION

$$
\begin{array}{l} \quad\qquad\qquad H \\ \text{— — — — — — — } CH_3-\underset{\underset{OH}{|}}{C}-CH_2-CO-S-ACP \end{array} \begin{array}{l} HS-Cys \\ \end{array}\!\!\!\Big\rangle E \quad \textit{D-3-hydroxybutyryl-ACP}
$$

$$
H_2O \!\!\! \Big\rangle \quad [\text{3-ketoacyl-ACP-dehydratase}]
$$

$$
\begin{array}{l} \quad\qquad\qquad H \\ \text{— — — — — — — } CH_3-C=\underset{\underset{H}{|}}{C}-CO-S-ACP \end{array} \begin{array}{l} HS-Cys \\ \end{array}\!\!\!\Big\rangle E \quad \textit{crotonyl-ACP}
$$

SATURATION $H^+ + NADPH$ ⟩ [2,3-*trans*-enoylacyl-ACP-reductase]

$$
\text{— — — — — — —} \begin{array}{l} HS-Cys \\ CH_3-CH_2-CH_2-CO-S-ACP \end{array} \!\!\! \Big\rangle E \quad \textit{butyryl-ACP}
$$

Figure 9.4 The biosynthesis of fatty acids (E = fatty acid synthase complex, ACP = acyl carrier protein).

THE ADDITION OF TWO-CARBON UNITS

The next metabolic step is the condensation of the acetyl and malonyl groups to form an acetoacetyl group attached to the ACP with the liberation of carbon dioxide. It may seem odd to add carbon dioxide and then lose it a few moments later but the energy needed to build the fatty acid chain from acetyl groups is supplied by this carboxylation–decarboxylation. The remaining reactions involving reduction, dehydration and saturation all take place on the ACP (Fig. 9.4).

The last enzyme catalyses the transfer of the butyryl group to the —SH group of the cysteinyl residue and the sequence is repeated with the chain lengthened by another two carbons. This is repeated for a total of seven cycles to form the 16-carbon palmitoyl group, which in micro-organisms is transferred to CoA to form palmitoyl-CoA and in animals is liberated as free palmitate by the action of a *deacylase*.

THE OVERALL REACTION

The biosynthesis of palmitate from acetyl-CoA can therefore be summarized:

$$8CH_3CO—S—CoA + 7ATP + 14NADPH + 14H^+$$

$$\downarrow$$

$$CH_3(CH_2)_{14}COO^- + 7ADP + 7P_i + 14NADP^+ + 6H_2O$$

Modification of fatty acids

ELONGATION AND UNSATURATION

Palmitate is now modified by lengthening of the hydrocarbon chain and the introduction of double bonds.

Elongation The 16-carbon chain of palmitate can be lengthened by a series of reactions on the endoplasmic reticulum that are similar to those of the fatty acid synthetase complex. Palmitoyl-CoA is formed from palmitate and two carbon units are added from malonyl-CoA after decarboxylation. The microsomal elongation system acts on unsaturated as well as on saturated fatty acids.

The chain can also be lengthened by an enzyme system found in mitochondria in which two carbon units are added as acetyl-CoA to the fattyacyl-CoA. The reactions are the reverse of β-oxidation and involve cleavage, reduction, dehydration and reduction. The enzymes and cofactors for this elongation are also similar to those of β-oxidation, apart from the last reductive step which is catalysed by enoyl reductase and uses NADPH as the coenzyme rather than $FADH_2$.

Unsaturation The unsaturated fatty acids are synthesized by oxidation and desaturation of palmitoyl-CoA and other fattyacyl-CoAs. The enzyme complex is present in the mitochondria and requires molecular oxygen, NADPH and cytochrome b_5 as cofactors. Mammals cannot introduce double bonds into the fatty acid chain beyond C_9, which is why linoleate ($18:2^{\Delta\,9,\,12}$) and linolenate ($18:3^{\Delta\,9,\,12,\,15}$) are essential fatty acids and must be supplied in the diet.

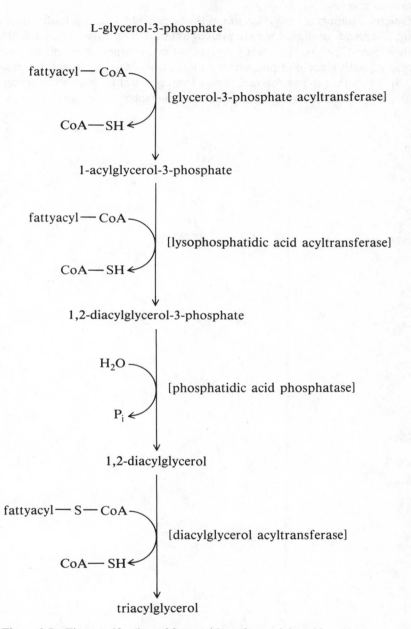

Figure 9.5   The esterification of fatty acids to form triglycerides.

ESTERIFICATION

One final important modification of the fatty acids is their esterification, as nearly all the fatty acids in organisms are present as esters. The fatty acids are first converted to their acyl-CoA derivatives by fattyacyl-CoA synthetase in microsomes and then reacted with glycerol-3-phosphate to form one, two and then three ester links (Fig. 9.5). The glycerol-3-phosphate comes from glycerol by phosphorylation or from the glycolytic intermediate dihydroxyacetone phosphate (Fig. 8.8).

9.3 Cholesterol: its metabolism and role in heart disease

Cholesterol is widespread throughout the animal kingdom and high concentrations are found in liver, brain and egg yolk. The sterol is an important component of membranes and occupies a key role in metabolism (Fig. 9.6) where it is the precursor of all the steroids synthesized by the body including bile acids, corticosteroids and the sex hormones (Section 3.5). The metabolism of cholesterol and its association with coronary heart disease is considered in this section and its conversion to the steroid hormones in Section 9.4.

Biosynthesis

Some of the cholesterol metabolized and excreted by the body comes from foods of animal origin but most of it is synthesized by the body.

ACETYL-COA TO ISOPENTENYL PYROPHOSPHATE

Our knowledge of the metabolic pathway for the biosynthesis of cholesterol arose from the work of a number of biochemists, principally Cornforth, Popjak, Bloch and Lynen. Studies with isotopes showed that all the carbon atoms of cholesterol come from acetyl-CoA which is converted to 3-hydroxy-3-methylglutaryl-CoA (HMG-CoA) by the pathway shown in Fig. 8.13. HMG-CoA is reduced, phosphorylated and decarboxylated to form isopentenyl pyrophosphate (Fig. 9.7) and like many other pathways, the first reaction of the metabolic sequence is the rate-limiting step.

ISOPENTENYL PYROPHOSPHATE TO CHOLESTEROL AND ITS ESTERS

The steroid skeleton is formed by the condensation of six of these five-carbon isoprene

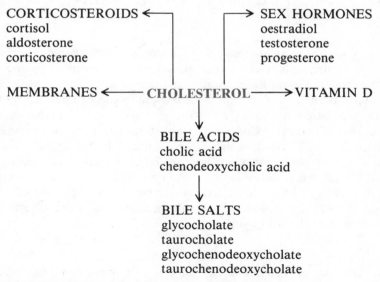

Figure 9.6 The central role of cholesterol in metabolism.

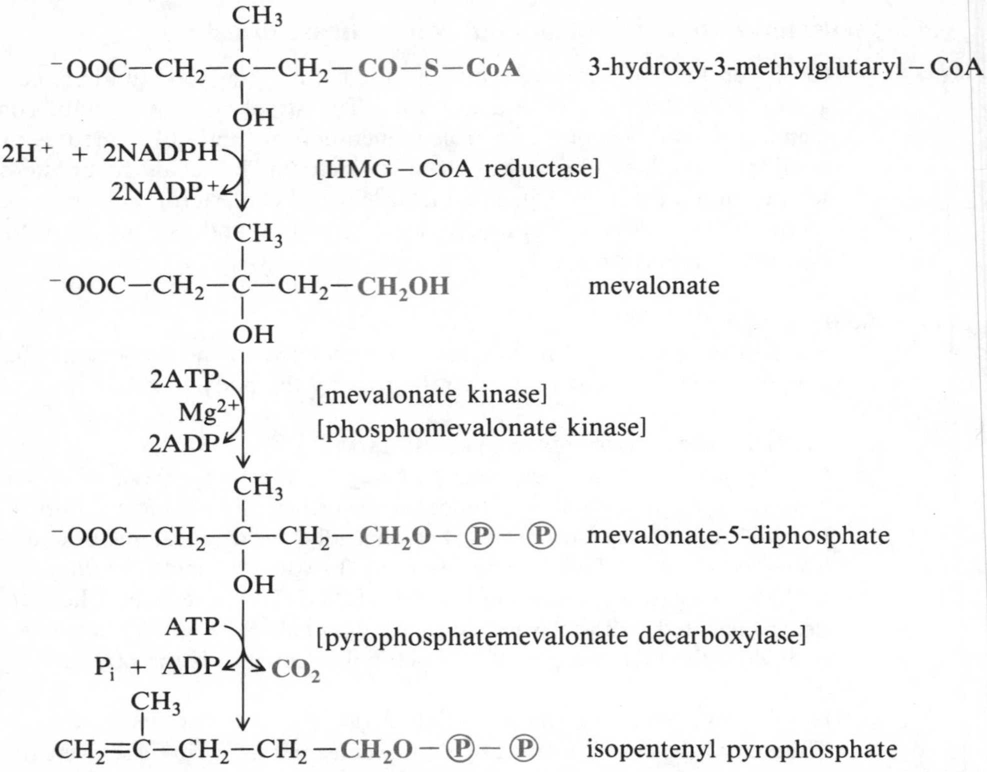

$$CH_3$$
$$^-OOC-CH_2-\overset{|}{\underset{|}{C}}-CH_2-CO-S-CoA \qquad \text{3-hydroxy-3-methylglutaryl} - CoA$$
$$OH$$

$$2H^+ + 2NADPH$$
$$2NADP^+ \qquad \text{[HMG} - \text{CoA reductase]}$$

$$CH_3$$
$$^-OOC-CH_2-\overset{|}{\underset{|}{C}}-CH_2-CH_2OH \qquad \text{mevalonate}$$
$$OH$$

$$2ATP$$
$$Mg^{2+} \qquad \text{[mevalonate kinase]}$$
$$2ADP \qquad \text{[phosphomevalonate kinase]}$$

$$CH_3$$
$$^-OOC-CH_2-\overset{|}{\underset{|}{C}}-CH_2-CH_2O-\textcircled{P}-\textcircled{P} \qquad \text{mevalonate-5-diphosphate}$$
$$OH$$

$$ATP$$
$$P_i + ADP \qquad CO_2 \qquad \text{[pyrophosphatemevalonate decarboxylase]}$$
$$CH_3$$
$$CH_2=\overset{|}{C}-CH_2-CH_2-CH_2O-\textcircled{P}-\textcircled{P} \qquad \text{isopentenyl pyrophosphate}$$

Figure 9.7 The synthesis of isopentenyl pyrophosphate from 3-hydroxy-3-methylglutaryl-CoA.

units and a simplified version of this pathway is shown in Fig. 9.8. Incidentally, isopentenyl pyrophosphate is also the precursor of a number of important plant products including the carotenes, rubber and camphor.

Cholesterol is an alcohol and the hydroxyl group can be esterified by long-chain fatty acids. The reaction is catalysed by the enzyme *cholesterol acyltransferase* which acts on cholesterol and acyl-CoA to give cholesterol esters and coenzyme A. Another enzyme, *cholesterol esterase*, which is also present in the cytosol, catalyses the hydrolysis of cholesterol esters to cholesterol and fatty acids and the concentration of cholesterol in the cell is controlled by the action of these two enzymes. Most of the cholesterol and its esters are incorporated into membranes or converted to bile salts (Section 3.5) and the remainder is used to synthesize the steroid hormones in the adrenal cortex and the sex organs.

Transport

Cholesterol is a very hydrophobic molecule and is transported in the blood as part of the plasma lipoproteins (Table 9.1). There are four main groups of these lipoprotein

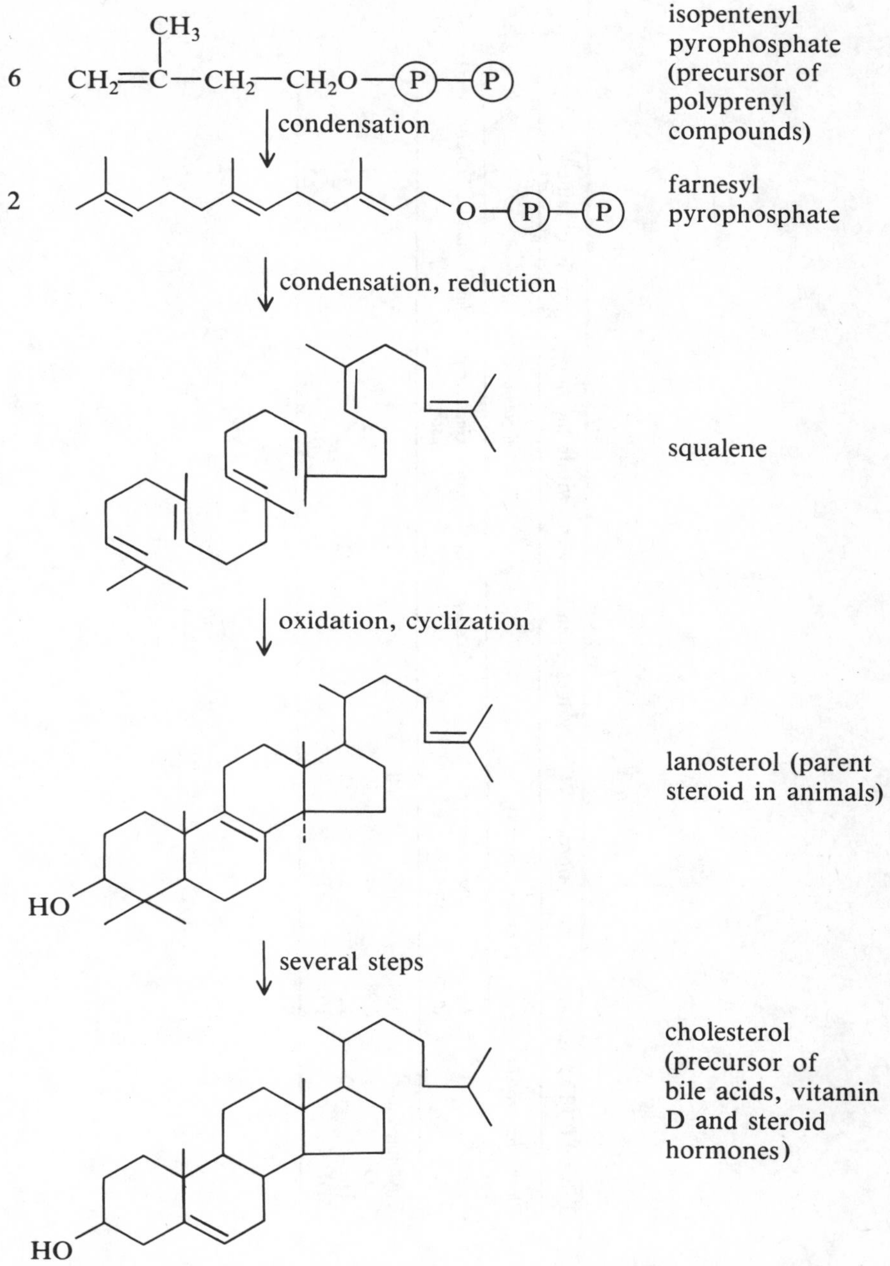

Figure 9.8 A greatly simplified metabolic pathway showing the synthesis of cholesterol from isopentenyl pyrophosphate.

Table 9.1 The human plasma lipoproteins (The values given below are within the range normally obtained)

Plasma lipoprotein	Physical property			Chemical composition (%)				
	Density (g cm^{-3})	Diameter (nm)	MW (×10^6)	Triglyceride	Cholesterol	Cholesterol ester	Phospholipid	Protein
Chylomicrons	0.94	500	500	87	2	3	7	1
VLDL	0.98	50	10	53	7	14	17	9
LDL	1.04	22	3	13	12	25	25	25
HDL	1.14	11	0.3	4	4	16	24	52

particles which are classified according to their density. These are chylomicrons, very low density lipoproteins (VLDL), low density lipoproteins (LDL) and high density lipoproteins (HDL) and each of these plays a part in the transport of cholesterol.

The following description applies to lipoprotein metabolism in man which is different to that of the laboratory rat.

CHYLOMICRONS

These particles are formed after a fatty meal and carry triglyceride from the intestine to the tissues of the body (Section 8.3). They also transport dietary cholesterol and take up cholesterol esters following their degradation by lipoprotein lipase. The degraded chylomicrons then deliver cholesterol and its esters to the liver.

VERY LOW DENSITY LIPOPROTEINS

These plasma lipoproteins transport triglyceride synthesized by the liver and also the intestine to other tissues of the body. As they circulate in the blood, fatty acids are removed in the tissues by lipoprotein lipase and the triglyceride content falls. At the same time the particles take up cholesterol and its esters and eventually become low density lipoproteins (LDL).

LOW DENSITY LIPOPROTEINS

The low density lipoproteins bind to receptors on the surface of cells and are taken up by lysosomes where they are degraded with the liberation of cholesterol and cholesterol esters. These lipoproteins therefore deliver free cholesterol and its esters to the tissues of the body.

HIGH DENSITY LIPOPROTEINS

As the chylomicrons and the VLDL are reduced in size, HDL are split off their surface and take up cholesterol. This is one source of HDL but most is synthesized and secreted by the liver as disc-shaped particles which mature into globular structures as they take up cholesterol from the blood. The cholesterol is then esterified by cholesterol acyltransferase and the resulting cholesterol ester forms the hydrophobic core of the particles. High density lipoproteins are therefore a reservoir of cholesterol and carry the sterol from the tissues to the liver.

Excretion

TURNOVER OF THE BILE

The circulation of cholesterol in the body occurs not only via the plasma lipoproteins but also through the turnover of the bile. Nearly all the bile secreted into the intestine is recovered but some of the bile acids and cholesterol are not reabsorbed. Cholesterol is the precursor of the bile acids (Section 3.5) so the major route whereby cholesterol or its products are lost from the body is in the faeces. This loss must be replaced and most of the cholesterol that is absorbed or synthesized is used for this purpose.

GALLSTONES

Cholesterol is an important component of the bile and when this is concentrated in the gall bladder the cholesterol can become supersaturated and precipitate to form gall stones. The precise cause of this is not known but biliary stasis and infection play a part. This condition is by no means rare and is three times more common in women than men. The usual treatment is to remove the gall bladder by surgery but drugs are being developed that lower the concentration of cholesterol in the bile by inhibiting the synthesis of the sterol.

Cholesterol and coronary heart disease

Cardiovascular disease is the single biggest cause of death in the United Kingdom and many other developed countries, particularly tragic are the premature deaths in relatively young adults from heart attacks. There are a number of factors that increase the risk of a heart attack (Table 9.2) and anyone who is unfortunate enough to have all the risk factors should change his or her lifestyle if he or she wishes to draw the old age pension.

Table 9.2 Risk factors for heart attacks

Factor	Increased risk	Decreased risk
Cholesterol	High	Low
Lipoproteins	Raised LDL	Raised HDL
Blood pressure	High	Normal
Weight	Overweight	Correct weight
Exercise	None and infrequent	Moderate and regular
Menopause	Post-menopausal	Pre-menopausal
Coffee	High intake	Low intake
Smoking	Smoker	Non-smoker
Alcohol intake	None or high	Moderate
Stress	High stress	Contentment

BLOOD CHOLESTEROL

The risk factor that has received most publicity is a raised blood cholesterol and many studies, including a long term survey in the town of Framingham in the United States, have shown a clear correlation between morbidity and plasma cholesterol. The normal range of plasma cholesterol is from 3.5 to 6.5 mM and in Framingham the group with a cholesterol greater than the top end of this range had about four times as many deaths from heart disease as those with a low plasma cholesterol. Cholesterol has been singled out for special attention but other fats are also raised in the blood of those most at risk.

Most cardiovascular disease starts very young with *atheroma*, when fatty deposits containing a high concentration of cholesterol and cholesterol esters are laid down in the walls of arteries. These atheromatous plaques of fatty materials later become fibrous and calcified and the atheroma develops into *atherosclerosis* or hardening of the arteries. These deposits narrow the lumen of the artery and provide a rough surface for a blood clot to form in a stroke or coronary thrombosis.

CHOLESTEROL LIPOPROTEINS

The blood cholesterol is a good statistical indicator of the risk of a heart attack but the level of the plasma lipoproteins appears to be more useful in this respect. People with a high level of LDLs have an increased risk of coronary heart disease while those with a high level of HDLs have a decreased risk. This is because LDLs are taking cholesterol to the tissues while HDLs are removing cholesterol from the tissues and the arterial walls. Moderate exercise taken regularly lowers the LDLs and raises the HDLs and sensible exercise is known to reduce the risk of a heart attack.

OTHER RISK FACTORS OF CORONARY HEART DISEASE

Feeding experimental animals a diet rich in cholesterol raises the concentration in the blood and increases the formation of fatty deposits in the arteries. Other studies in humans have shown that fats rich in saturated fatty acids raise the blood cholesterol while those rich in polyunsaturated fatty acids lower the blood cholesterol. In order therefore to reduce the risk of a heart attack, the intake of eggs rich in cholesterol and animal fat that contains mainly saturated fatty acids should be reduced and replaced by vegetable oils and fish that are rich in polyunsaturated fatty acids.

Diet is not the only or even perhaps the most important factor contributing to arterial disease as the 'Irish brothers' study has shown. The health of two groups of Irish brothers was followed over 10 years; one brother had emigrated to Boston in the United States while the other brother had stayed in Ireland. Post-mortem examination of the hearts of those who died accidentally showed that the 'Irish' hearts were much healthier than the 'American' hearts. Furthermore, those that had remained in Ireland had a lower blood pressure, a lower plasma cholesterol and were generally more healthy than their American counterparts. This was in spite of the higher intake of meat and dairy products by the brothers in Ireland. Other less tangible factors have therefore to be considered such as contentment and lack of stress.

9.4   The steroid hormones: their synthesis and biological action

The steroid hormones which are synthesized by the adrenal cortex and the sex organs are all derived from cholesterol. The major and minor sources of the different types of hormones are shown in Table 9.3 and summaries of their biological activities are given in Section 3.5.

Table 9.3   The steroid hormones

Type and example	No. of C atoms	Major source	Minor source
Oestrogens Oestrone Oestradiol	18	Ovary, placenta	Adrenal cortex, testis
Progestins Progesterone	21	Corpus luteum, placenta	Adrenal cortex, testis
Androgens Testosterone	19	Testis	Adrenal cortex, ovary
Corticosteroids Aldosterone Cortisol	21	Adrenal cortex	—

Biosynthesis

SYNTHESIS OF PREGNENOLONE

The first steps in the synthesis of all the steroid hormones take place in the mitochondria when cholesterol is converted to *pregnenolone* by hydroxylation and the removal of six of the carbons from the hydrocarbon side chain at C_{17} (Fig. 9.9).

METABOLISM OF PREGNENOLONE

Pregnenolone is the major precursor of all the steroid hormones and a scheme for the pathways of *steroidogenesis* is shown in Fig. 9.10. This scheme has been simplified since only the most important of the steroid hormones are considered and not all the synthetic pathways are shown. Some of the routes are more important than others and the main pathway taken for the synthesis of a particular hormone depends in some cases on the tissue where the synthesis takes place. For example, the oestrogens can be synthesized via androstenediol or via progesterone and testosterone. In the case of the human female, the former pathway is the more important in the Graafian follicle and the latter in the corpus luteum of pregnancy.

The scheme in Fig. 9.10 clearly shows the importance of hydroxylations in the biosynthesis of the steroid hormones and the enzymes catalysing these reactions are all *hydroxylases* or *mixed function oxidases* and require oxygen, NADPH and cytochrome P450 (Section 10.4). The location of these and the other enzymes of steroidogenesis depends on the reaction catalysed but most transformations take place in the mitochondria and on the endoplasmic reticulum.

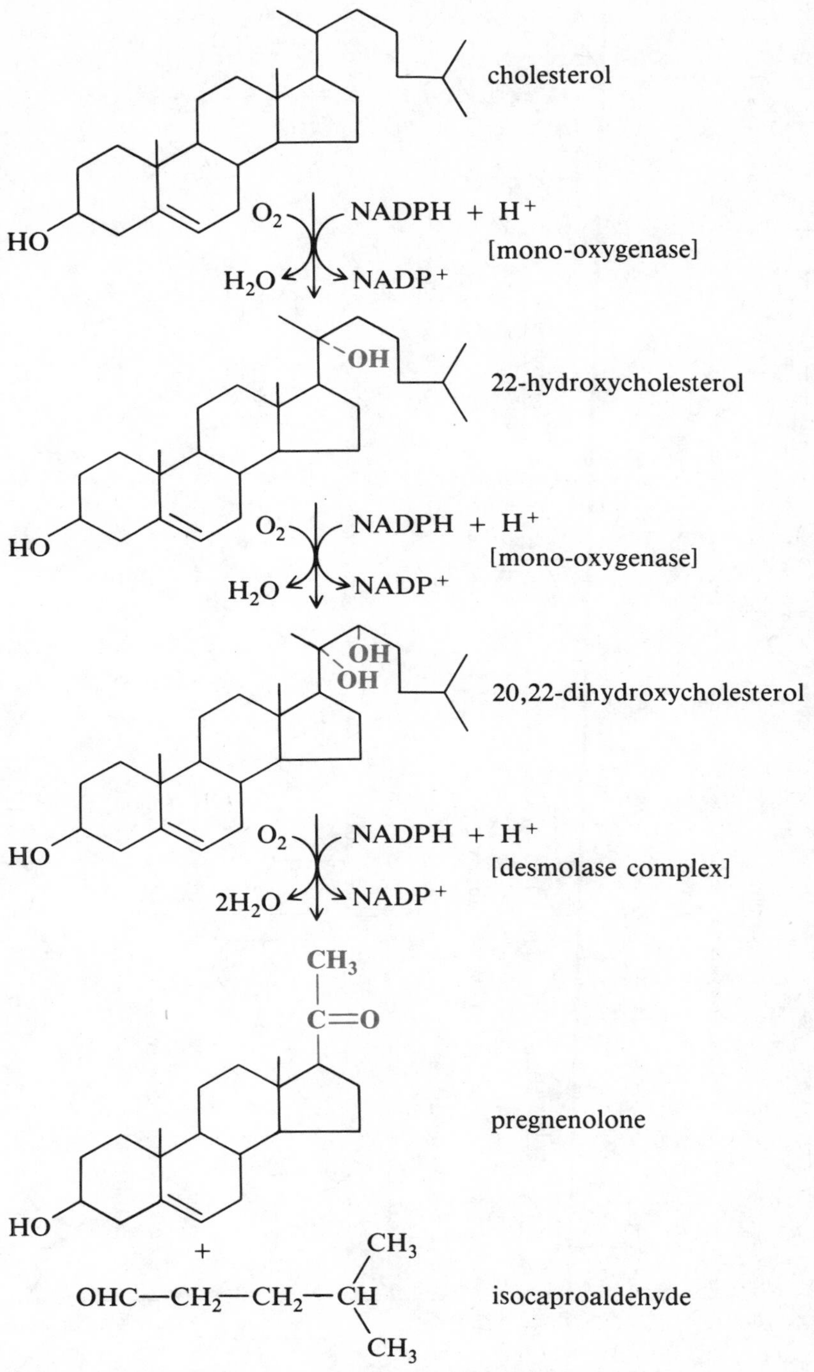

cholesterol

O_2 NADPH + H$^+$

[mono-oxygenase]

H_2O NADP$^+$

22-hydroxycholesterol

O_2 NADPH + H$^+$

[mono-oxygenase]

H_2O NADP$^+$

20,22-dihydroxycholesterol

O_2 NADPH + H$^+$

[desmolase complex]

$2H_2O$ NADP$^+$

pregnenolone

+

$OHC-CH_2-CH_2-CH \begin{smallmatrix} CH_3 \\ \\ CH_3 \end{smallmatrix}$ isocaproaldehyde

Figure 9.9 The conversion of cholesterol to pregnenolone.

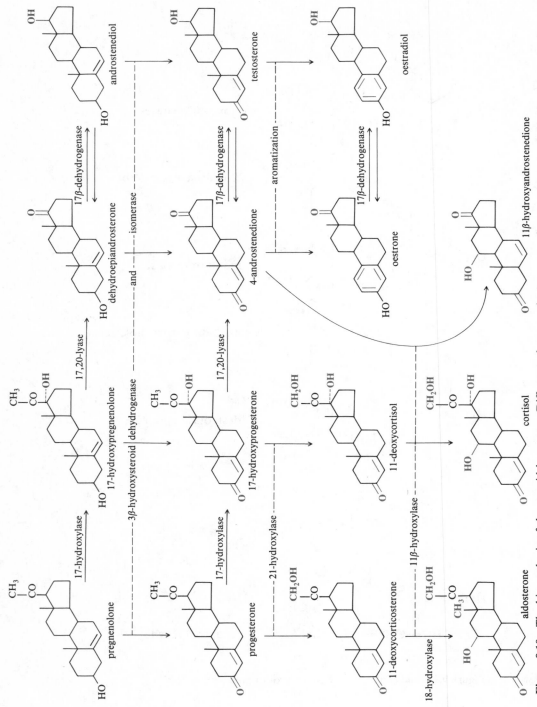

Figure 9.10 The biosynthesis of the steroid hormones. Differences between the structure of pregnenolone and the hormones are shown in red.

Biological action

OESTROGENS

The oestrogens, named after their ability to induce *oestrus* in immature female rats, are unusual in that part of the steroid skeleton is an aromatic ring. They are synthesized by the ovaries and by the placenta during pregnancy. The oestrogens are responsible for the secondary sexual characteristics of the female and together with progesterone they control and regulate the menstrual cycle and pregnancy. *Oestradiol* and *oestrone* are the main oestrogens synthesized by the ovaries and *oestriol* is the major oestrogen produced by the placenta during pregnancy. Oestriol has the same structure as oestradiol with an extra —OH group at position 16. The most potent of these hormones is oestradiol and if the biological activity of oestriol is taken to be 1 then oestrone is 2.5 and oestradiol 25.

PROGESTINS

The only natural progestin is *progesterone* although there are a number of synthetic analogues with progestational activity. Progesterone is synthesized mainly by the corpus luteum during pregnancy and in small quantities by the adrenal cortex and the testis as an intermediate in the synthesis of other steroid hormones (Fig. 9.10). Its function is to prepare the uterus to receive the fertilized ovum and to maintain pregnancy. It does this by stimulating the growth of specific target organs including the uterine mucosa, the maternal placenta and the mammary glands.

ANDROGENS

The androgen with the highest biological activity is testosterone secreted by the interstitial cells of the testis. It is the major sex hormone in males and its *androgenic activity* is responsible for the development and maintenance of the male secondary sexual characteristics. Testosterone also has *anabolic activity* and stimulates growth, particularly of bone and muscle. Androgens are also produced by the adrenal cortex but the activity of these steroids is much less than testosterone.

CORTICOSTEROIDS

The adrenal cortical steroids are essential for life and any major deviation from the normal output causes widespread metabolic and physiological disturbances. This occurs whether the cortical steroid production is too little as in *Addison's disease* or too much as in *Cushing's syndrome*. Both of these conditions are serious and treatment consists of bringing the circulating level of the hormones back within the normal range.

There are two types of corticosteroids although there is some overlap of activity (Table 9.4). The *mineralocorticoids* are produced by the outer and smallest zone of the cortex and the *glucocorticoids* by the two inner zones which also produce small amounts of androgens and oestrogens. *Aldosterone*, the dominant mineralocorticoid, is a powerful hormone which is involved in water and electrolyte homoeostasis. It acts on the renal tubules and promotes the absorption of Na^+ and the excretion of K^+.

Cortisol and *corticosterone* are glucocorticoids which have widespread effects on the metabolism of carbohydrates, fats and proteins. They stimulate glycogenesis in the liver and also gluconeogenesis which leads to a rise in the blood sugar. The glucocorticoids mobilize fatty acids in adipose tissue although they are not lipolytic themselves but act permissively in allowing the response to glucagon, adrenaline, etc. They also increase the breakdown of protein to give a negative nitrogen balance, promote water diuresis and maintain the normal blood pressure.

Table 9.4   The relative biological activities of some natural and synthetic steroids

Steroid	Glucocorticoid activity (glycogen deposition)	Mineralocorticoid activity (Na^+ retention)
Natural steroids		
Cortisol	1	1
Corticosterone	0.1	15
Aldosterone	0.3	600
Synthetic steroids		
Prednisolone	5	0.8
9-α-Fluorocortisol	10	125
Dexamethasone	25	0

Synthetic steroids

A small change in the structure of a steroid can have quite a profound effect on its biological activity. For example, replacing the CH_3—CO— side chain at position 17 by —OH converts the female sex hormone progesterone to the male sex hormone testosterone. Slight changes in structure can also enhance the normal physiological effect of a hormone (Table 9.4) and several synthetic analogues have been developed as drugs (Fig. 9.11).

ORAL CONTRACEPTIVES

Synthetic oestrogens and progestins similar to those illustrated (Fig. 9.11) are used in oral contraceptives to prevent pregnancy. They act by inhibiting ovulation and preventing the implantation of a fertilized ovum in the uterus. There are however unfortunate side effects and the most serious of these is the development of a thrombus which is fatal in a small number of women.

ANABOLIC STEROIDS

Methandienone is an anabolic steroid which increases muscle mass and strength and is sometimes taken illegally by athletes in the field events of the shot, hammer, discus and javelin. Detection of anabolic steroids and their metabolites in urine is one way that reputable sporting authorities are seeking to curb their abuse. Prolonged use of such drugs may win an athlete a medal but are also likely to make him or her sterile and prone to serious disease such as diabetes, heart disease and cancer.

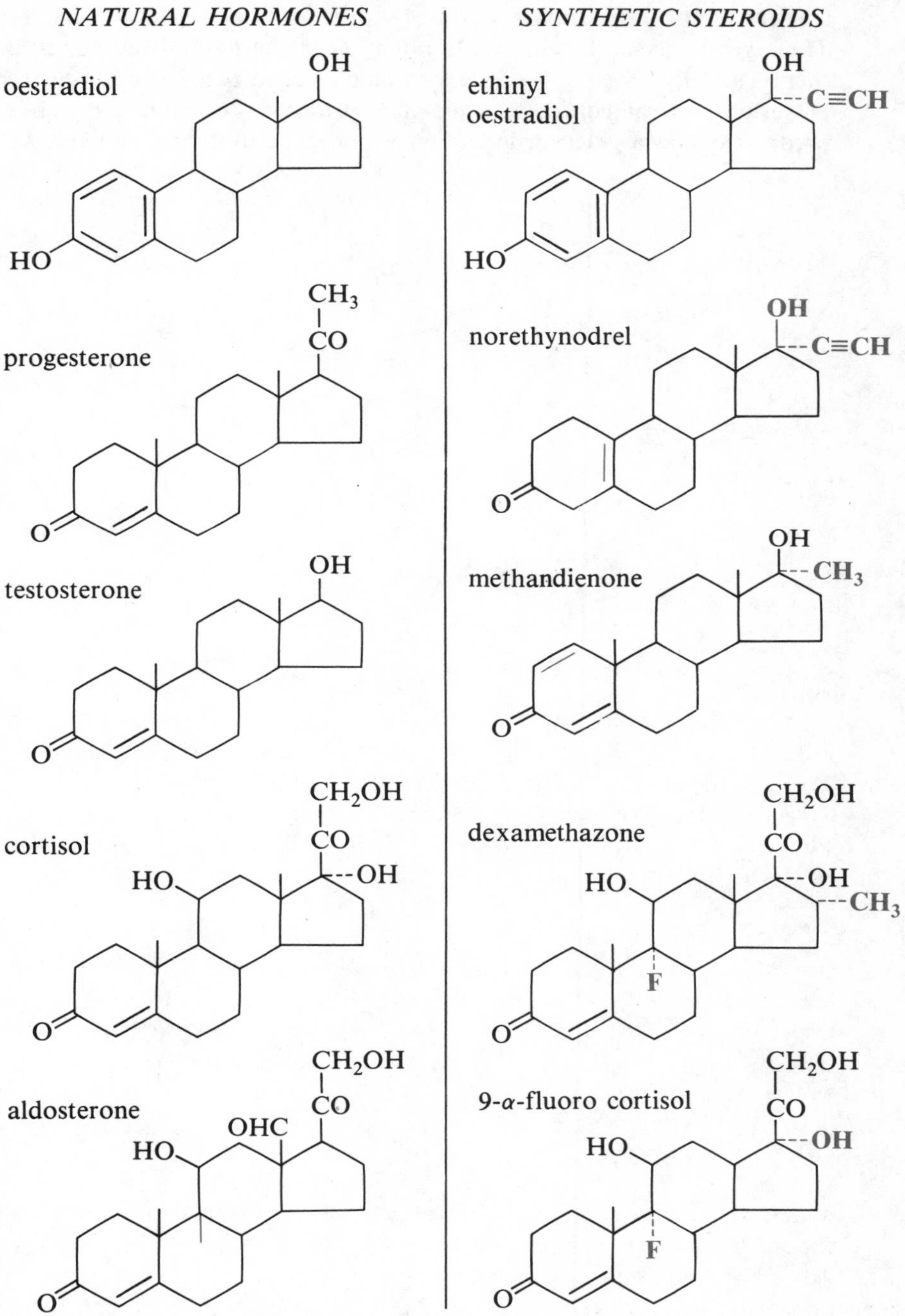

NATURAL HORMONES

SYNTHETIC STEROIDS

oestradiol

ethinyl
oestradiol

progesterone

norethynodrel

testosterone

methandienone

cortisol

dexamethazone

aldosterone

9-α-fluoro cortisol

Figure 9.11 The structure of some natural and synthetic steroids.

SYNTHETIC CORTICOSTEROIDS

These synthetic steroids are more potent than the natural hormones and are used therapeutically. *Prednisolone* is a synthetic glucocorticoid and *9-α-fluorocortisol* is a powerful mineralocorticoid while *dexamethasone* is even more powerful than prednisolone as a glucocorticoid but has no mineralocorticoid activity (Table 9.4).

9.5   Amino acids: precursors of active compounds

As discussed earlier, amino acids in an organism are in a dynamic state of equilibrium (Fig. 8.17). Many are used to synthesize proteins and those surplus to requirements are deaminated and their carbon skeletons oxidized or converted to carbohydrates or fats. In addition to this general metabolism, amino acids are precursors of many important compounds. For example, aspartate, glycine and glutamine are needed for the synthesis of *purines* and aspartate for the production of *pyrimidines*. Glycine and taurine derived from cysteine are essential components of *bile acids* (Section 3.5), while methionine, threonine and glycine are required for the synthesis of *vitamin B_{12}*. Some idea of the diversity of this can be seen in Fig. 9.12 which shows the possible

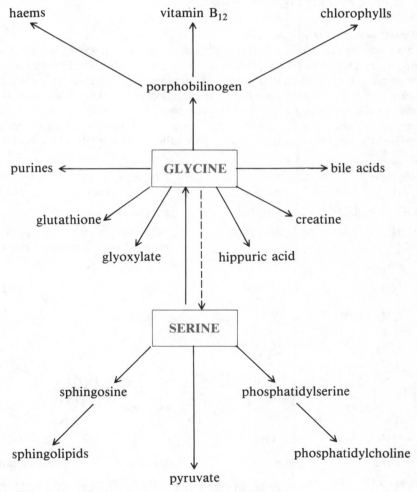

Figure 9.12   The metabolism of glycine and serine.

metabolic fates of glycine and serine. The breadth of this subject can also be judged from the fact that each one of the 22 amino acids normally found in proteins is involved in several metabolic pathways. This section can therefore only consider selected aspects of specialized amino acid metabolism and the examples chosen are all from human biochemistry and have a pharmacological or medical interest.

Biogenic amines

The decarboxylation of amino acids or their metabolites gives rise to amines, some of which are pharmacologically very active. These amines are often present at low concentrations but they can produce quite marked changes in mammalian physiology. The compounds selected to illustrate this point are neurotransmitters but also show other pharmacological activities.

DOPAMINE AND NORADRENALINE

These amines are produced in the adrenal medulla or nervous tissue from tyrosine (Fig. 9.13). Dopamine is a powerful vasoconstrictor but is best known for its action in the nervous system as an inhibitory neurotransmitter. In *Parkinson's disease* there is a deficiency of this compound due to neuronal degeneration: this affects the balance between the excitatory and inhibitory transmitters and gives rise to the uncontrolled tremors which are a symptom of this disease. Treatment for this condition is to increase the dopamine in the brain by giving the drug L-DOPA (3,4-dihydroxyphenyl-

Enzymes			*Cofactors*
[TH]	=	tyrosine hydroxylase	tetrahydrobiopterin and NADPH
[DDC]	=	DOPA decarboxylase	pyridoxal phosphate
[DH]	=	dopamine hydroxylase	ascorbate, a co-substrate, is reduced to dehydroascorbate

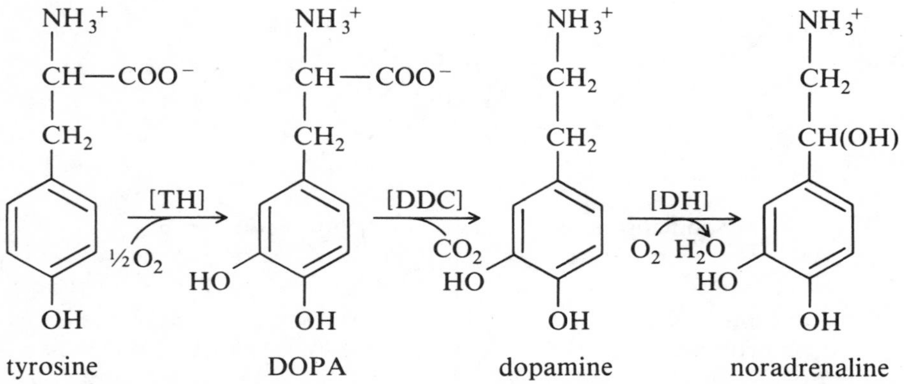

Figure 9.13 The production of the neurotransmitters dopamine and noradrenaline from tyrosine.

alanine). This is the amino acid precursor of the neurotransmitter and is converted to dopamine by decarboxylation (Fig. 9.13).

Over-production of dopamine can also be a problem and causes psychological disturbances with symptoms of *schizophrenia*. Treatment in this case is to give a drug such as *chlorpromazine* which blocks the dopamine receptor.

Noradrenaline, which is derived from dopamine (Fig. 9.13), is another neurotransmitter found in the brain and the sympathetic nervous system. It is also synthesized in the adrenal medulla and together with adrenaline enables the animal to deal with emergency situations.

SEROTONIN

5-Hydroxytryptamine (serotonin) is synthesized from tryptophan by hydroxylation and decarboxylation using enzymes that are similar to those involved in the formation of dopamine from tyrosine (Fig. 9.13).

$$HO\text{—}\underset{N}{\boxed{}}\text{—}CH_2\text{—}CH_2\text{—}NH_3{}^+ \qquad \text{5-hydroxytryptamine (serotonin)}$$

Serotonin is another pharmacologically active compound which is a neurotransmitter in the central nervous system, stimulates the contraction of smooth muscle and is a powerful vasoconstrictor. Serotonin is also an important precursor of other compounds and when acetylated and methylated forms *melatonin* (methoxy-*N*-acetyltryptophan), the pineal hormone responsible for circadian rhythms. Serotonin is also found in plants as a precursor of hallucinogenic compounds present in some species of mushrooms.

HISTAMINE

Histamine is formed by decarboxylation of histidine.

$$HN\text{—}\underset{}{\boxed{}}\text{—}N\text{—}CH_2\text{—}CH_2\text{—}NH_3{}^+$$

histamine

The amine is a neurotransmitter particularly in the brains of insects but it is probably best known for its involvement in the allergic response. Stress stimulates histamine decarboxylase in mast cells and histamine is released at the site of injury; this is particularly noticeable in the skin following an insect bite.

Histamine also stimulates acid secretion in the stomach and drugs such as *cimetidine*, which are used to treat peptic ulcers, act by blocking histamine receptors and thereby reducing the acidity of the stomach.

Inborn errors of metabolism

Some individuals are born with a defective gene so that a particular enzyme has a low activity or is not synthesized at all. If this happens, then the end product of the metabolic pathway will not be produced and the metabolite normally converted by the enzyme will accumulate in the cell. There are many examples of these inborn errors in the metabolism of amino acids and three conditions involving defective enzymes of tyrosine metabolism are now briefly considered (Fig. 9.14).

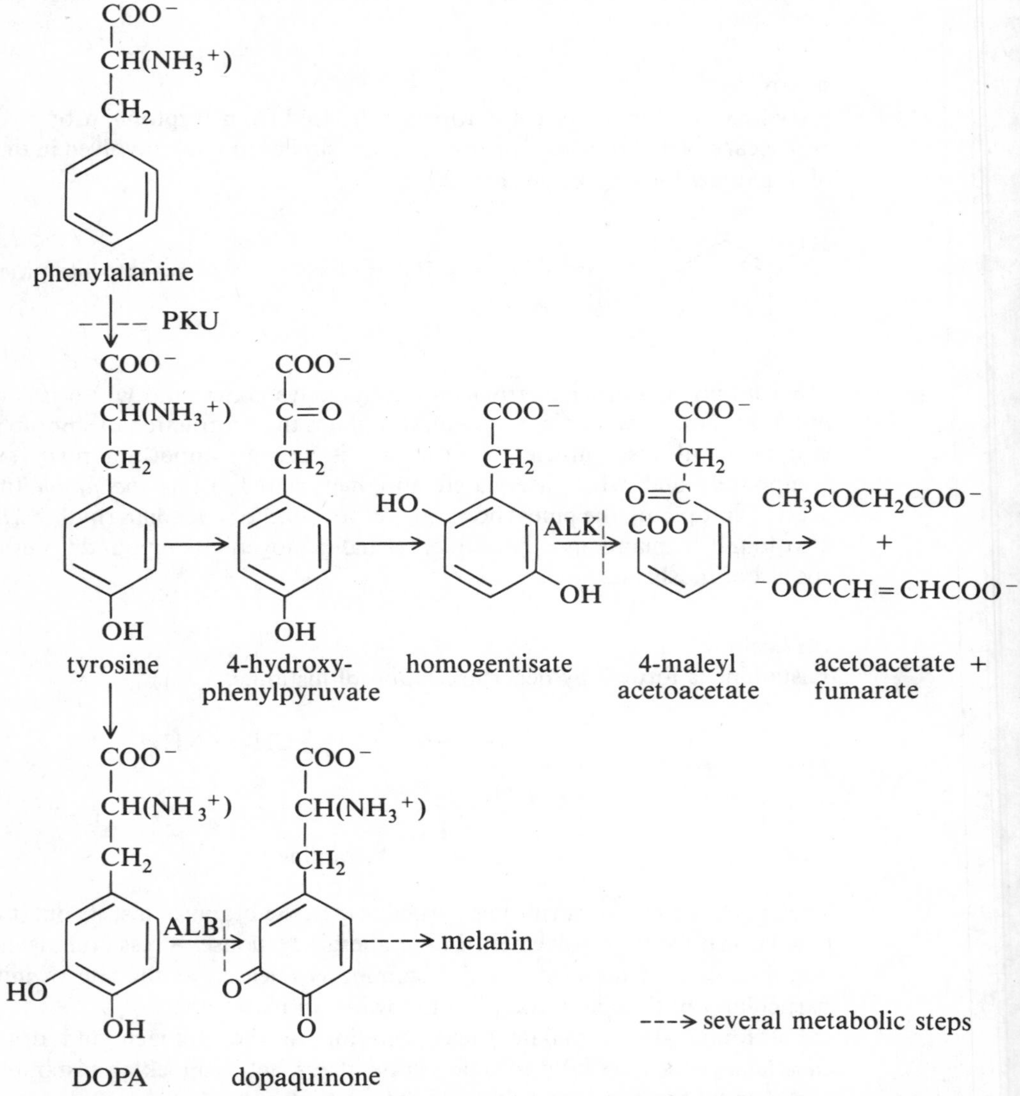

Figure 9.14 The missing or defective enzymes (shown by red lines) in a number of inborn errors of metabolism (PKU = phenylketonuria, ALB = albinism, ALK = alkaptonuria).

PHENYLKETONURIA

The missing enzyme in phenylketonuria is *phenylalanine hydroxylase* which converts phenylalanine to tyrosine:

$$O_2 + \text{phenylalanine} \qquad \text{tetrahydrobiopterin} \qquad NADP^+$$

$$H_2O + \text{tyrosine} \qquad \text{dihydrobiopterin} \qquad NADPH + H^+$$

phenylalanine hydroxylase

dihydrobiopterin reductase

The formation of tyrosine is thus blocked and the phenylalanine is directed down what would normally be minor metabolic pathways (Fig. 9.15). There is therefore a greatly increased excretion of the minor metabolite phenylpyruvate in the urine and it is this that gives the disease its name. The excess circulating levels of phenylpyruvate and other metabolites interfere with the synthesis of myelin in the brain and this is the probable cause of the mental retardation associated with the untreated condition. The mental retardation can be prevented if the condition is diagnosed soon after birth and the infant placed on a diet with a restricted intake of phenylalanine. The diet is expensive but still considerably cheaper than the cost of keeping someone in a mental institution for life.

ALBINISM

Melanin is the dark pigment responsible for the colour of skin, hair and eyes. It is synthesized from tyrosine via DOPA, DOPA quinone and a series of complex polymerization reactions. Albinos are unable to synthesize melanin as they lack the enzyme *tyrosinase* which converts tyrosine to DOPA and DOPA to DOPA quinone (Fig. 9.14). This condition is not uncommon in laboratory animals which exhibit the characteristic features of white fur, pale skin and pink eyes. It is less common in man and the majority of cases are partial with patches of unpigmented skin and hair.

ALKAPTONURIA

This condition, studied by Garrod early this century, led to the concept of genetic disease and inborn errors of metabolism. The characteristic feature is the passing of urine which turns black on standing. The melanin-like pigment is formed from *homogentisate* which is a normal product of tyrosine metabolism but which accumulates in large amounts in this condition because of a missing enzyme, *homogentisate oxidase* (Fig. 9.14). There are few other problems associated with this inborn error apart from the rather spectacular symptom of producing a black urine and a proneness to arthritis later in life.

Hormones

Amino acids are essential components of a number of polypeptide hormones such as vasopressin (Fig. 3.4), oxytocin, glucagon and insulin and the substitution of only one

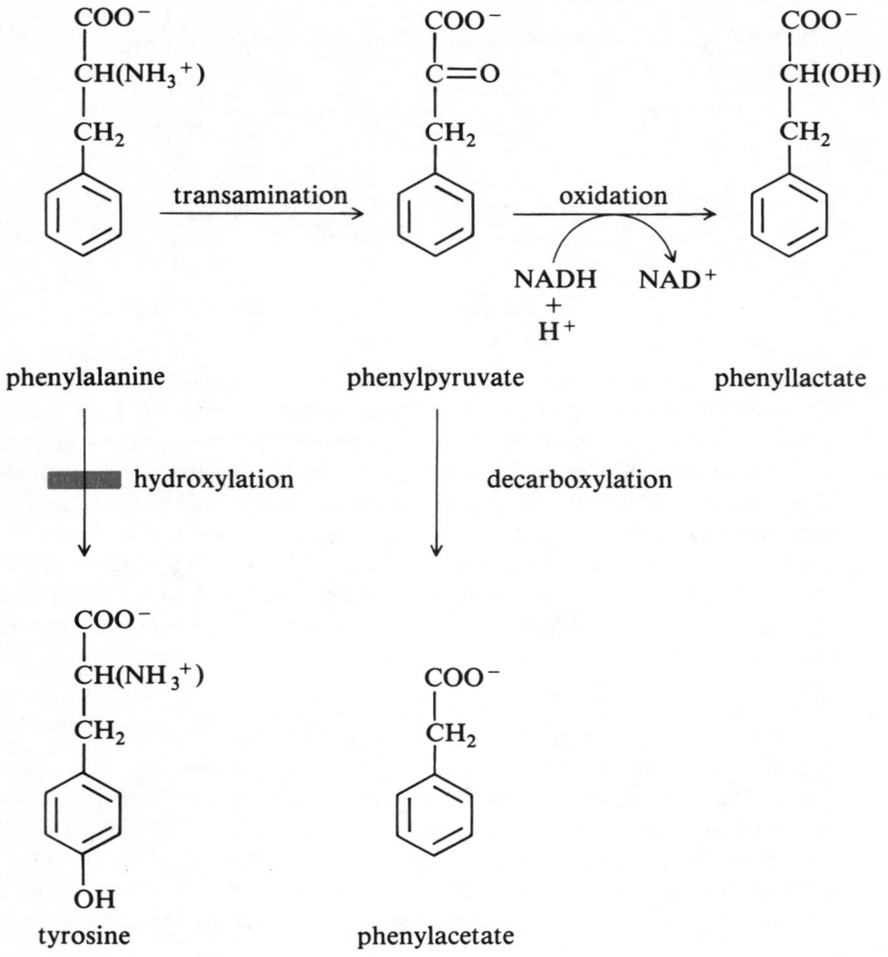

(■■■■ represents blockage due to missing or defective enzyme)

Figure 9.15 The enhancement of the minor pathways of phenylalanine metabolism in phenylketonuria.

amino acid can have a profound effect on the hormonal activity. However, the examples selected are not peptides but metabolites of just one amino acid tyrosine.

THYROID HORMONES

Tyrosine residues are iodinated and converted to the thyroid hormones of which the most active are triiodothyronine and thyroxine:

triiodothyronine and thyroxine (extra I)

The hormones act by stimulating the energy metabolism in the tissues of the body. The exact mechanism of this is not clear but is probably related to their effect on the swelling of mitochondria and the uncoupling of oxidative phosphorylation. Iodine is concentrated by the thyroid gland and reacts with tyrosyl residues of the protein *thyroglobulin* to form mono- and di-iodinated products. Triiodothyronine comes from the coupling of these iodinated tyrosyl residues and thyroxine from the condensation of two molecules of diiodotyrosine. The free hormone is liberated from thyroglobulin by lysosomal proteases and carried in the blood to the target cells bound to a thyroid-binding globulin.

There are a number of pathological conditions associated with the over- or under-production of the thyroid hormones. An increased synthesis causes *thyrotoxicosis* which is characterized by a raised basal metabolic rate (BMR) and increased nervous energy. A reduced synthesis causes *myxoedema* in adults which has the opposite symptoms of a lowered BMR, sluggishness and mental apathy. The persistence of a low concentration of thyroid hormones in infants is called *cretinism* the main features of which are stunted growth and mental retardation. Another condition associated with a defective thyroid is *goitre* where there is an enlargement of the thyroid gland which in most cases is due to a deficiency of iodine in the diet. It is particularly common in areas such as Derbyshire which have a thin, iodine-deficient soil. Goitre caused by lack of iodine can be easily prevented by adding iodide to salt or bread.

ADRENALINE

Adrenaline is synthesized in the adrenal medulla by noradrenaline-*N*-methyltransferase which catalyses the methylation of noradrenaline by *S*-adenosylmethionine.

adrenaline (epinephrine)

Adrenaline, together with noradrenaline, is released from the adrenal gland in response to emotional stress such as anger, pain or fright. These two hormones are chemically amines of the phenol catechol and are often referred to as *catecholamines*. Together they prepare the animal to deal with dangerous situations by escaping or resisting the threat.

Together they produce a series of complex but integrated changes in the physiology and biochemistry of the animal. Blood is diverted away from the skin and the gut and towards the skeletal muscles; glucose and fat are mobilized and there is a marked increase in respiration. Thus a fright leads to flight or fight!

9.6 Nitrogen fixation: a vital process in the biosphere

The importance of nitrogen fixation

Nitrogen is an essential component of many important biomolecules such as proteins, porphyrins and nucleic acids and the circulation of this element between the atmosphere and the biosphere in the nitrogen cycle is extremely important (Section 2.3). A vital part of this process (Fig. 2.8) is nitrogen fixation whereby molecular nitrogen is converted to a form which can be assimilated and used by living organisms.

A small amount of nitrogen is fixed in nature during thunderstorms, when nitrogen is oxidized along the path of a lightning discharge, but most fixation takes place in micro-organisms by reduction of the element to ammonia. Nitrogen is also fixed industrially by reduction and is introduced into the biosphere through the application of fertilizers.

THE INDUSTRIAL FIXATION OF NITROGEN

A large amount of nitrogen is fixed in industry in the Haber process by reduction of the element to ammonia.

$$3H_2 + N_2 \longrightarrow 2NH_3 \qquad \Delta G^0 = -33 \, \text{kJ}$$

The overall reaction is energetically favourable with a large negative G^0 but the energy consumed by this process, which operates at high temperatures and pressures, is high. This is because a large input of energy is needed to obtain hydrogen from the methane of natural gas and also to break the $N \equiv N$ bond which has a high bond energy of $940 \, \text{kJ mol}^{-1}$.

A small amount of nitrogen is fixed by oxidation, mostly inadvertently. Some of this is a by-product of industry but most comes from the exhaust fumes of the ubiquitous internal combustion engine.

REDUCTION OF NITROGEN BY MICRO-ORGANISMS

Nitrogen fixation in the biosphere occurs only in certain micro-organisms and is not found in the higher forms of life. Nitrogen fixation in the oceans of the world is carried out by blue-green algae, while that on land takes place in micro-organisms which are free living in the soil or in symbiotic association with fungi and plants. Nitrogen fixation in the free-living organisms such as *Azotobacter* has been the most studied but the biggest contribution comes from symbiotic organisms such as *Rhizobia* which are found in the root nodules of legumes such as clover, peas and broad beans.

Nitrogenase

ENZYME STRUCTURE

The key enzyme in nitrogen fixation is *nitrogenase* which has been purified from a number of species and has a rather complicated structure. The enzyme complex contains iron and molybdenum and both metals are needed for enzyme activity.

Nitrogenase is made up of two iron–sulphur (Fe—S) or non-haem iron proteins. In these structures, the iron is not part of a haem centre as in haemoglobin and the cytochromes but is associated with sulphur in the form of an Fe—S cluster (Fig. 9.16). The structure of the enzyme varies slightly with the source but there are always two components and both are essential for enzyme activity. The smallest of the ferroproteins has a molecular weight of about 60 000 and consists of two identical subunits. The largest ferroprotein, with a molecular weight in the region of 220 000, is more complicated and is made up of two types of subunits each one of which is a dimer. It is this largest protein that contains the molybdenum which is probably present as part of an Mo—Fe—S cluster.

The reduction of N_2 to NH_3 requires six electrons:

$$N_2 + 6e^- + 6H^+ \longrightarrow 2NH_3$$

but nitrogenase actually uses eight electrons and produces hydrogen as a by-product

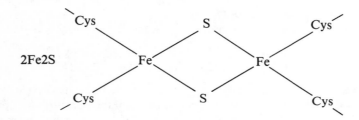

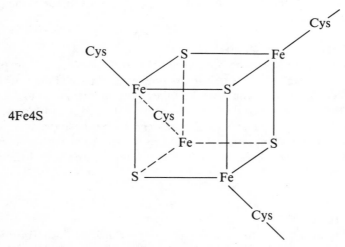

Figure 9.16 Structures of iron–sulphur clusters present in non-haem iron proteins (Cys = cysteine residues of the protein linked to the iron via the sulphur atom).

so the equation for the enzyme catalysed reaction is:

$$N_2 + 8e^- + 8H^+ \longrightarrow 2NH_3 + H_2$$

Reducing power is provided by electrons flowing via an electron transport chain to *ferredoxin* or *flavodoxin*, depending on the species, so that anything reducing ferredoxin (including light in photosynthesis) can supply electrons for nitrogen fixation. Two molecules of ATP are consumed for each electron transferred so taking into account the ATP, the reaction catalysed by nitrogenase is:

$$N_2 + 8e^- + 8H^+ + 16ATP \longrightarrow 2NH_3 + H_2 + 16ADP + 16P_i$$

The fixation of nitrogen is therefore energetically costly since each molecule of N_2 requires 16 molecules of ATP.

Nitrogenase also catalyses the reduction of other compounds (Table 9.5) and the production of ethylene from acetylene can be detected by gas–liquid chromatography (GLC) and used to assay the enzyme activity.

Table 9.5 The reduction of substrates by nitrogenase

Substrate	Product
N_2	NH_3
H^+	H_2
C_2H_2	C_2H_4
N_2O	N_2
CN^-	$CH_4 + NH_3$

SENSITIVITY TO OXYGEN

Oxygen is toxic to nitrogenase and the enzyme loses its activity in air in a matter of only a few minutes. Various strategies have therefore been developed in nature to overcome the problem of oxygen toxicity. These include synthesizing more nitrogenase in the presence of oxygen, forming a barrier between oxygen and the enzyme or lowering the concentration of oxygen by rapidly removing it. A good example of this last mechanism is seen in the case of *Rhizobia* in the root nodules of legumes which manufacture a form of haemoglobin known as *leghaemoglobin*. This is a very neat way of solving the problem of how to protect the nitrogenase from oxygen yet at the same time supplying enough oxygen for the cytochrome oxidase to function. Leghaemoglobin effectively buffers free oxygen to levels below those toxic to nitrogenase, but leaves enough oxygen available for the cytochrome oxidase.

Nitrate and nitrite

The ammonia produced by nitrogen fixation is taken up by plants and used for nitrogen metabolism. The major route of assimilation is via glutamine synthetase, glutamate synthetase and transaminase. However, the nitrogen in the soil is best present as nitrate, since this is more readily taken up by the plants than ammonia, and nitrate is produced from ammonia by the combined action of two species of soil bacteria. *Nitrosomonas* contains the enzyme *nitrite reductase* which catalyses the oxidation of ammonia to nitrite:

$$NH_4^+ + 2H_2O \longrightarrow NO_2^- + 8H^+ + 6e^-$$

Protons are also produced and electrons which are passed on to a cytochrome electron transport chain. *Nitrobacter* then completes the oxidation when *nitrate reductase* catalyses the conversion of nitrite to nitrate:

$$NO_2^- + H_2O \longrightarrow NO_3^- + 2H^+ + 2e^-$$

Protons are again produced and the electrons passed on to an electron transport chain containing cytochromes.

Similar enzymes are found in plants but these catalyse the reverse reactions so that nitrate taken up from the soil can be converted to ammonia which is then used for the synthesis of amino acids.

Nitrate and nitrite are used by denitrifying bacteria as the terminal electron acceptor instead of oxygen. Nitrogen gas is produced which then passes into the atmosphere. This loss from the biosphere appears to be wasteful but denitrification is an essential process as it replenishes the atmospheric nitrogen which would otherwise become depleted.

Nitrate is used as a fertilizer and some of this is reduced to nitrite by soil bacteria. Nitrates and nitrites are highly soluble compounds that are easily washed out of the soil by rain and find their way into the water supply and this can be a serious source of environmental pollution. Nitrites are the real problem since they react with haemoglobin to form the oxidized product *methaemoglobin* which is unable to carry oxygen. Furthermore, nitrites form nitrous acid in the acid environment of the stomach and this reacts with secondary amines in the food to form *nitrosamines* which are highly carcinogenic:

$$
\begin{array}{ccc}
R_1 & & R_1 \\
\diagdown & & \diagdown \\
\quad NH + HNO_2 \longrightarrow & & \quad N{-}N{=}O + H_2O \\
\diagup & & \diagup \\
R_2 & & R_2 \\
\text{amine} & & \text{nitrosamine}
\end{array}
$$

Future developments

BIOTECHNOLOGY

The Haber process for the fixation of nitrogen has served man for the best part of a century but it consumes large quantities of energy (Section 2.3). It also suffers from the disadvantage that the hydrogen is obtained from methane present in natural gas which is a non-renewable resource. Biological nitrogen fixation on the other hand takes place at ambient temperature and pressure. It would be of great benefit to develop an industrial process based on that found in nature by using catalysts containing molybdenum or related metals and using the renewable energy of the sun.

GENETIC ENGINEERING

Another area being investigated is the isolation of the gene coding for nitrogenase and its incorporation into the genomes of other micro-organisms and even plants that do not have the enzyme. This would be particularly valuable in growing crops on poor soils with a low nitrogen content.

ENZYME INHIBITION

Denitrification is an important process globally to maintain the nitrogen balance of our planet but locally the loss of nitrogen can be wasteful. This is particularly so following the application of fertilizer when as much as 50 per cent of the nitrogen may be lost by denitrification. The development of a compound which would inhibit the denitrifying nitrate reductase and nitrite reductase and prevent this loss would be extremely useful. However, the compound must not inhibit the assimilatory enzymes in plants, which work in the opposite direction, and so far this problem has not been overcome.

10. Inactivation and detoxication

10.1   Ammonia: an essential but toxic by-product of metabolism

Detoxication of ammonia

TOXICITY OF AMMONIA

Quite large amounts of ammonia are produced as a by-product of the metabolism of amino acids and other nitrogenous compounds (Fig. 8.17) but ammonia is highly toxic to the organism. The precise mechanism of toxicity is not clear but it is generally thought that NH_3 rather than NH_4^+ is the toxic entity. The equilibrium favours the ionized form and at pH 7.4 only 1 per cent of the ammonia is present as NH_3:

$$NH_3 + H^+ \rightleftharpoons NH_4^+$$

However, this is enough to interfere with metabolism and toxic effects are seen when the blood concentration reaches 0.2 mM. The NH_3 readily penetrates membranes and an increased concentration in the tissues will shift the equilibrium of reactions involving this metabolite. For example, glutamate dehydrogenase will be shifted towards glutamate formation rather than breakdown, with a reduction in the 2-oxoglutarate, and the loss of this Krebs' cycle intermediate will in turn affect the oxidation of glucose and fatty acids. An increased concentration of ammonia also inhibits glutaminase which causes a fall in glutamate. These effects are most serious in the brain which is dependent on glucose oxidation and which uses glutamate and its metabolite γ-aminobutyric acid (GABA) as neurotransmitters.

TEMPORARY STORAGE OF AMMONIA

Some of the ammonia reacts with 2-oxoglutarate and NADH or NADPH to form *glutamate* in a reaction catalysed by glutamate dehydrogenase. The glutamate can

$$
\begin{array}{c}
\text{COO}^- \\
|\\
\text{CH}_2 \\
|\\
\text{CH}_2 \\
|\\
\text{CH(NH}_3^+) \\
|\\
\text{COO}^-
\end{array}
\;+\; NH_4^+ \;+\; ATP \;\xrightarrow[\text{Mg}^{2+}]{\text{[GS]}}\;
\begin{array}{c}
\text{CONH}_2 \\
|\\
\text{CH}_2 \\
|\\
\text{CH}_2 \\
|\\
\text{CH(NH}_3^+) \\
|\\
\text{COO}^-
\end{array}
\;+\; ADP \;+\; P_i \;+\; H_2O
$$

glutamate   glutamine

then be used to synthesize amino acids by transamination (Section 8.6). Ammonia can also be removed by reaction with the acidic amino acids to form the corresponding amides *glutamine* and *asparagine*. These provide a temporary store of ammonia and are formed from glutamate and aspartate by the action of *glutamine synthase* (GS) and *asparagine synthase* respectively.

Glutamine is important in animals where its formation in peripheral tissues enables ammonia to be transported to the liver in a non-toxic form. The amide is also a nitrogen donor in the synthesis of purines and pyrimidines and is needed for rapidly dividing cells such as lymphocytes and the intestinal mucosa. The ammonia stored in glutamine is released by the action of *glutaminase* (G) which catalyses the hydrolysis of glutamine to glutamate and ammonia:

$$
\begin{array}{c}
CONH_2 \\
| \\
CH_2 \\
| \\
CH_2 \\
| \\
CH(NH_3^+) \\
| \\
COO^-
\end{array}
\quad + \ H_2O \ \xrightarrow{[G]} \quad
\begin{array}{c}
COO^- \\
| \\
CH_2 \\
| \\
CH_2 \\
| \\
CH(NH_3^+) \\
| \\
COO^-
\end{array}
\quad + \ NH_4^+
$$

glutamine   glutamate

This reaction is important in the liver where it is the main source of the ammonia used in the urea cycle. Glutaminase is also very active in the kidney where the ammonia released is used to protect the body against acidosis.

EXCRETION PRODUCTS

Glutamate and glutamine are only temporary stores of ammonia and the waste product needs to be removed completely. There are various ways that animals do this depending on the availability of water (Table 10.1).

Table 10.1 The major nitrogenous excretion products of animals

Compound excreted	Water availability	Type of animals
Ammonia	Plenty	*Ammoniotelic animals* Aquatic vertebrates and amphibian larvae, e.g. tadpoles
Urea	Limited	*Ureotelic animals* Terrestrial vertebrates including man, e.g. adult frogs
Uric acid	More limited	*Uricotelic animals* Terrestrial invertebrates birds and reptiles, e.g. bats

Ammonia is toxic but can be excreted directly by bony fish into a large volume of water where it is diluted to a harmless concentration.

Urea is excreted by most land animals who have less access to water. It is an ideal waste product as it has a high nitrogen content (47 per cent w/w), is very soluble indeed (6 M at 25 °C) and is non-toxic.

Uric acid is excreted as an insoluble slurry of crystals by animals in the desert where the availability of water is even more limited.

In human beings some ammonia is excreted directly into the urine but the major excretion product is urea which can be as much as 30 g per day depending on the amount of protein in the diet. The conversion of ammonia into urea takes place in the liver in the urea cycle.

The urea cycle

DISCOVERY

The experiments that led to the discovery of the urea cycle were carried out by Krebs and Henseleit in 1932. These workers showed that the formation of urea by liver slices in a buffered ionic medium was stimulated by the amino acids ornithine, citrulline and arginine. These amino acids form a sequence in which two molecules of ammonia are incorporated into one molecule of urea:

$$
\begin{array}{cccc}
 & NH_2 & NH & CO(NH_2)_2 \\
 & | & | & \\
 & C{=}O & C{=}NH_2{}^+ & + \\
 & | & | & \\
NH_3{}^+ \quad NH_4{}^+ & NH \quad NH_4{}^+ & NH \quad H_2O & NH_3{}^+ \\
| & | & | & | \\
HCO_3{}^- \; + \; (CH_2)_3 \longrightarrow & (CH_2)_3 \longrightarrow & (CH_2)_3 \longrightarrow & (CH_2)_3 \\
| \qquad\qquad\quad & | \qquad\quad & | \qquad & | \\
CH(NH_3{}^+) \; 2H_2O & CH(NH_3{}^+) \; H_2O & CH(NH_3{}^+) & CH(NH_3{}^+) \\
| & | & | & | \\
COO^- & COO^- & COO^- & COO^-
\end{array}
$$

| ornithine | citrulline | arginine | urea + ornithine |

The ornithine is regenerated and so the sequence of reactions forms a metabolic cycle and the urea cycle, as originally proposed by Krebs and Henseleit, is shown in Fig. 10.1.

STAGES IN THE UREA CYCLE

The advent of radioisotopes enabled the details of the cycle to be worked out and a more modern version of the urea cycle is given in Fig. 10.2. The cycle obviously functions as a complete entity but for ease of learning this is broken down into three stages.

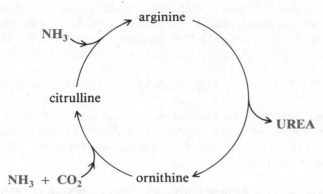

Figure 10.1 The urea cycle as originally proposed by Krebs and Henseleit.

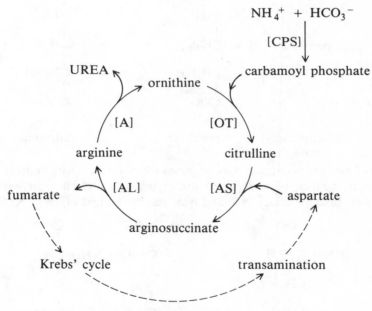

[CPS] = carbamoyl phosphate synthase
[OT] = ornithine transcarbamoylase
[AS] = arginosuccinate synthase
[AL] = arginosuccinate lyase
[A]   = arginase

Figure 10.2 The urea cycle as understood today.

1. *Synthesis of carbamoyl phosphate* Ammonia from the action of glutaminase and glutamate dehydrogenase reacts with carbon dioxide in the form of bicarbonate and ATP to give carbamoyl phosphate. This reaction is catalysed by the enzyme *carbamoyl phosphate synthase* (CPS) and takes place in liver mitochondria:

$$HCO_3^- + 2ATP + NH_4^+ \xrightarrow[Mg^{2+}]{[CPS]} H_2N—COO\textcircled{P} + 2ADP + P_i + H_2O + H^+$$

As with the Krebs' cycle this first reaction of the urea cycle is probably the rate-limiting step and is stimulated by *N*-acetylglutamate. After a protein meal, glutamate is acetylated with acetyl-CoA and the *N*-acetylglutamate stimulates the activity of carbamoyl phosphate synthase.

2. *Formation of arginine* Carbamoyl phosphate then reacts with ornithine to form citrulline, a reaction catalysed by *ornithine transcarbamoylase* (OT):

carbamoyl   ornithine   citrulline
phosphate

The citrulline then passes into the cytoplasm where it reacts with another molecule of ammonia in the form of aspartate to give arginosuccinate. This reaction is catalysed by *arginosuccinate synthase* (AS) and is driven by the hydrolysis of ATP:

citrulline   aspartate   arginosuccinate

The last reaction in this part of the cycle is the removal of fumarate by *arginosuccinate lyase* (AL) to form arginine:

$$
\begin{array}{ccc}
\underset{\text{arginosuccinate}}{
\begin{array}{l}
NH_2 \\
| \\
C-O-NH-CH \\
| \qquad\qquad\quad | \\
NH_2 \qquad\quad CH_2 \\
| \qquad\qquad\quad | \\
(CH_2)_3 \qquad COO^- \\
| \\
CH(NH_3^+) \\
| \\
COO^-
\end{array}
}
&
\xrightarrow{\text{[AL]}}
&
\underset{\text{arginine}}{
\begin{array}{l}
NH_2 \\
| \\
C=NH_2^+ \\
| \\
NH_2 \\
| \\
(CH_2)_3 \\
| \\
CH(NH_3^+) \\
| \\
COO^-
\end{array}
}
\quad + \quad
\underset{\text{fumarate}}{
\begin{array}{l}
HC-COO^- \\
\| \\
{}^-OOC-CH
\end{array}
}
\end{array}
$$

arginosuccinate   arginine   fumarate

3. *Production of urea*   The final step of the cycle is the hydrolysis of arginine by *arginase* (A) to give urea, and ornithine which then enters the mitochondria to start another turn of the cycle:

$$
\underset{\text{arginine}}{
\begin{array}{l}
NH_2 \\
| \\
C=NH_2^+ \\
| \\
NH_2 \\
| \\
(CH_2)_3 \\
| \\
CH(NH_3^+) \\
| \\
COO^-
\end{array}
}
+ H_2O
\xrightarrow{\text{[A]}}
\underset{\text{ornithine}}{
\begin{array}{l}
NH_2 \\
| \\
(CH_2)_3 \\
| \\
CH(NH_3^+) \\
| \\
COO^-
\end{array}
}
+ \underset{\text{urea}}{CO(NH_2)_2}
$$

arginine   ornithine   urea

THE ENERGY COST OF AMMONIA DETOXICATION

The complete urea cycle is shown in Fig. 10.2. At first sight it appears to be very costly in terms of the energy used since three molecules of ATP are needed for one molecule of urea. Furthermore, since one of the ATP molecules is broken down to $AMP + PP_i$ the real energy cost is four 'high energy phosphate bonds' which is equivalent to four ATP molecules broken down to $4ADP + 4P_i$. However, in order to complete the sequence of reactions, the conversion of fumarate to aspartate by the Krebs' cycle and transamination also need to be taken into account (Fig. 10.2). When this is considered, then three molecules of ATP are generated by oxidative phosphorylation from the NADH formed when the malate is oxidized to oxaloacetate by malate dehydrogenase, so that the net cost for the synthesis of one molecule of urea is only one molecule of ATP. For a typical amino acid this is about 15 per cent of the energy available by oxidation.

Finally, the urea passes into the blood and is lost from the body by excretion in the urine. Some nitrogen is recovered by ruminants and the camel when urea is converted to amino acids by micro-organisms in the gastrointestinal tract.

Clinical aspects

The urea cycle plays a vital role in protecting the animal against ammonia toxicity and any major interference with the ability to synthesize and excrete urea can have serious and sometimes fatal consequences.

METABOLIC DISORDERS

There are reports of enzyme defects for all five of the urea-cycle enzymes but they are fortunately rare. The clinical features are similar since they all involve ammonia intoxication. There is no cure for these conditions but some alleviation is possible by keeping the blood ammonia level as low as practicable. This minimizes brain damage but these disorders are serious and any major enzyme deficiency is lethal.

LIVER FAILURE

The urea cycle only operates in the liver so severe liver disease such as hepatitis, cirrhosis, or poisoning by hepatotoxins such as carbon tetrachloride means that the urea cycle is unable to operate efficiently. This leads to ammonia retention, coma and eventually death if untreated.

CHRONIC RENAL FAILURE

Failure of the kidneys means that the urea is not excreted and the level rises markedly in the blood. This *uraemia* is accompanied by the accumulation of many other metabolites and the best treatment is to remove the toxic products by haemodialysis or to replace the defective kidney by a transplant. However, in the less severe cases, some alleviation is possible by using some of the treatments below.

TREATMENT OF AMMONIA TOXICITY

Treatment consists of minimizing the amount of ammonia produced by restricting the intake of protein and ensuring that the amino acids present are balanced, so they are used for protein synthesis with only minimal deamination. The oxo-acids of essential amino acids are sometimes given as they effectively mop up ammonia when they use glutamate during transamination. The oxo-acids cannot be synthesized in adequate amounts by the body so that their exogenous supply also allows formation of 'essential' amino acids (since —NH_2 is available in excess) for normal growth.

Lactulose can also be given together with an insoluble antibiotic. Ammonia is produced by micro-organisms in the gut and the antibiotic reduces the population of bacteria. The lactulose is a synthetic disaccharide which is fermented in the colon and produces acid which lowers the pH and makes ammonia and amines less likely to diffuse into the bloodstream.

None of these treatments is ideal but they do help.

10.2 Oxygen: life giving but toxic

Oxygen is needed by all forms of life, apart from a few micro-organisms that are able to function anaerobically, and oxygen deprivation rapidly leads to death. However, this essential element is also toxic and exposure to pure oxygen at a pressure of one atmosphere for any length of time causes cellular damage and eventually death. This toxicity is not due to molecular oxygen but to its highly reactive free radicals that are produced in small amounts in normal metabolism. Under normal circumstances, the rate of destruction of these damaging radicals is greater than their rate of formation but there are situations when the rate of production of these reactive forms of oxygen is too high for the natural defence mechanisms to deal with.

Oxygen free radicals

Free radicals are species with one or more unpaired electrons (shown as a superscript dot) and most are short lived and highly reactive although a few are stable.

FORMS OF OXYGEN

Molecular oxygen, known as *dioxygen*, is actually a stable diradical as it has two unpaired electrons which occupy different energy levels. Oxygen can form three series of ionic oxides by the addition of electrons to dioxygen and some of these forms are extremely reactive:

$$O_2^{\cdot\cdot} \xrightarrow{e^-} O_2^{-\cdot} \xrightarrow{e^-} O_2^{2-\cdot\cdot} \xrightarrow{2e^-} 2O^{2-}$$

Superoxide $(O_2^{-\cdot})$ Superoxide free radicals have one more electron than dioxygen and are formed by the action of ionizing radiation on molecular oxygen. They are highly reactive and are one of the causes of radiation damage to cells and tissues. Traces of superoxide also occur naturally and are formed in the cell by enzymes and by the non-enzymatic autoxidation of compounds such as thiols and quinones. Small amounts are also produced when oxygen binds to haemoglobin and myoglobin and superoxide may be the bound form of oxygen in these haem proteins.

Peroxide $(O_2^{2-\cdot\cdot})$ Peroxides contain two more electrons than dioxygen and are also very reactive. The undissociated form of the peroxide dianion is hydrogen peroxide which is produced during normal metabolism by the action of oxidases such as D-amino acid oxidase and monoamine oxidase.

Oxide (O^{2-}) Normal oxides have two more electrons than atomic oxygen which gives a stable octet of electrons. They are the most reduced of the ionic oxides and are also the most stable.

Hydroxyl radical $(HO\cdot)$ This radical is actually half a molecule of hydrogen peroxide and is extremely reactive. Hydroxyl radicals are formed when atomic radiation passes through aqueous solutions containing oxygen and when hydrogen peroxide reacts with superoxide or metals such as iron and copper:

$$H_2O_2 + O_2^{-\cdot} \longrightarrow O_2 + HO^- + HO\cdot$$

$$H_2O_2 + Fe^{2+} \longrightarrow Fe^{3+} + HO^- + HO\cdot$$

CELLULAR DAMAGE BY OXYGEN RADICALS

Free radicals cause widespread damage throughout the cell by attacking nucleic acids, proteins and lipids. Modification of the bases in DNA causes mutagenesis and carcinogenesis while oxidation of the —SH groups of proteins to —S—S— leads to loss of flexibility and biological activity. However, the most serious acute damage is the lipid peroxidation of the unsaturated fatty acids of membranes which can be severe because of a self-perpetuating chain reaction (Fig. 10.3). Peroxidation of the membrane lipids makes them less hydrophobic and this affects membrane fluidity, flexibility and transport.

One of the products of the free-radical cleavage of unsaturated fatty acids is *malondialdehyde*. This active aldehyde reacts with the —NH$_2$ groups of proteins to

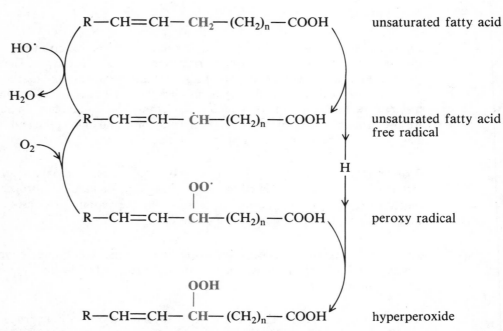

Figure 10.3 The chain reaction of lipid peroxidation triggered by free radicals.

form Schiff's bases and this is thought to be the origin of the insoluble pigment *lipofuschin* which accumulates in cells during ageing.

$$OHC-CH_2-CHO \text{ malondialdehyde}$$

Protection against oxygen toxicity

SUPEROXIDE DISMUTASE

Superoxide is rapidly removed by the metalloenzyme *superoxide dismutase* (SOD) which is present in the cytosol and in the mitochondria of all cells that use oxygen. Electrons are transferred from one molecule of superoxide to another to give hydrogen peroxide and molecular oxygen:

$$2H^+ + O_2^{-\cdot} + O_2^{-\cdot} \xrightarrow{\text{[SOD]}} H_2O_2 + O_2$$

The enzyme, which has a high capacity, is extremely efficient so that any superoxide is rapidly destroyed but there are some conditions when the production of superoxide exceeds the ability of SOD to remove it, with serious and often fatal results. An example of this is the herbicide *paraquat* which is highly toxic because of its ability to produce so much superoxide that the protective mechanism is unable to cope:

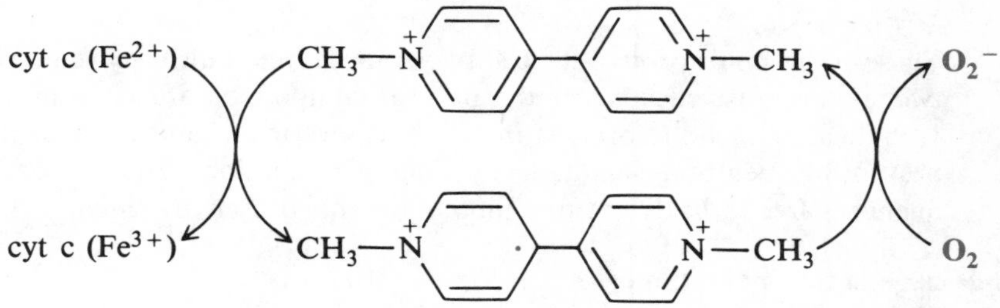

This reaction also gives rise to the highly reactive hydroxyl radicals when the superoxide or the paraquat free radical reacts with hydrogen peroxide.

CATALASE

The detoxication of hydrogen peroxide is carried out by the enzyme *catalase*. Hydrogen peroxide is removed very efficiently since catalase is present in the same organelle as the oxidases which produce it and the turnover number of the enzyme is very high. Catalase splits hydrogen peroxide into water and oxygen: the oxygen is liberated or used to oxidize a variety of compounds. In the latter case, catalase acts as a *peroxidase* and this reaction can serve a double purpose of detoxicating the other compound as well as hydrogen peroxide:

$$2H_2O_2 \xrightarrow{\text{catalase}} 2H_2O + O_2$$

$$H_2O_2 + CH_3CH_2OH \xrightarrow{\text{peroxidase}} 2H_2O + CH_3CHO$$

GLUTATHIONE PEROXIDASE

Organic peroxides and hydrogen peroxide are also removed by the action of the enzyme *glutathione peroxidase* (GP). This enzyme, which is found in the cytoplasm and the mitochondria, is unusual in that it contains the trace element selenium. Glutathione is a tripeptide of glutamate, cysteine and glycine in which the peptide bond between glutamate and cysteine is formed through the γ-carboxyl group of the acidic amino acid. The reduced form of the tripeptide is usually written as GSH to show the free —SH group of the cysteine residue:

γ-glutamyl-cysteinyl-glycine
glutathione (GSH)

The —SH group is susceptible to oxidation and the oxidized form of glutathione consists of two molecules covalently linked by a disulphide bond (GSSG). During the action of glutathione peroxidase, peroxide reacts with glutathione to give the oxidized form and water:

$$H_2O_2 + 2GSH \xrightarrow{[GP]} GSSG + 2H_2O$$

$$ROOH + 2GSH \xrightarrow{[GP]} GSSG + ROH + H_2O$$

ANTIOXIDANTS

These compounds protect the cell by acting as free-radical scavengers (Fig. 10.4) and two of the most effective of the natural compounds are vitamins. The fat-soluble vitamin E (α-tocopherol) acts in a lipid environment and is particularly effective at preventing membrane damage by lipid peroxidation. Vitamin C (ascorbate) also quenches free radicals but this time in an aqueous environment.

Oxygen damage in human erythrocytes

The highest concentration of oxygen is found in erythrocytes and this makes them susceptible to oxidation and damage by free radicals. Furthermore, in the case of mature human erythrocytes, there is no synthesis of new material so that damaged molecules and structures cannot be replaced or repaired. In this section we shall see how the metabolism in the human erythrocyte is used to defend the cell against oxygen toxicity and how a defective enzyme leads to the breakdown of this protection and causes haemolytic anaemia.

GENERAL METABOLISM OF THE ERYTHROCYTE

The mature human erythrocyte is a simple biconcave disc and is devoid of internal structures and organelles. There is no endoplasmic reticulum nor are there any mitochondria, lysosomes, peroxisomes or ribosomes so that the metabolic pathways associated wtih these structures are missing. The metabolism of the erythrocyte like its structure is also very simple and consists of glucose breakdown by glycolysis and the pentose phosphate pathway. These two pathways together provide the metabolites that are needed to maintain the functional integrity of the cell. For example, ATP is

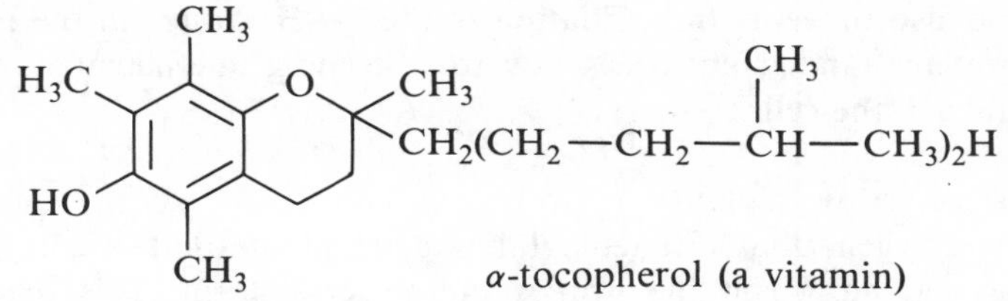

Figure 10.4 Some natural and synthetic antioxidants. They are all lipid soluble and remove free radicals by trapping them to form stable species.

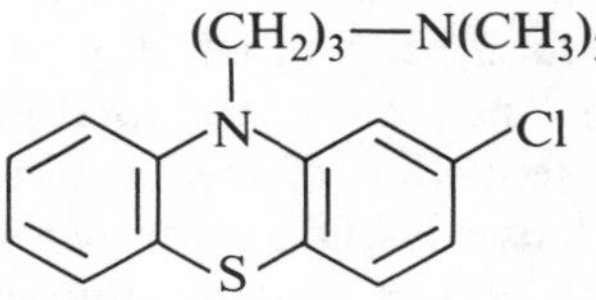

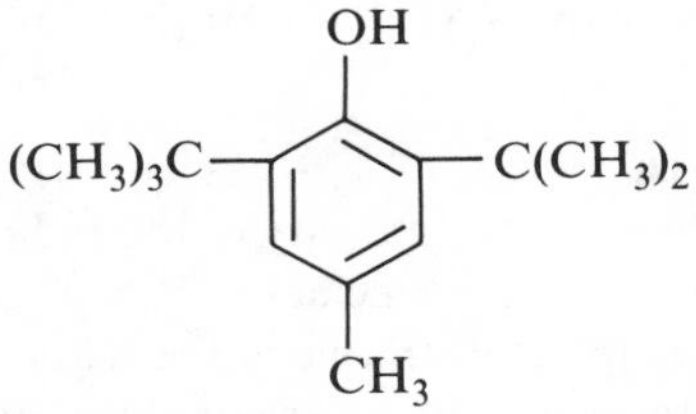

used to drive the Na^+ pump which maintains the ionic environment of the erythrocyte and 2,3-bisphosphoglycerate produced during glycolysis shifts the oxygen dissociation curve of haemoglobin into the physiological range of oxygen tension.

THE GENERATION OF GLUTATHIONE

Another important metabolite is glutathione. This is kept in the reduced state by the enzyme glutathione reductase (GR) which uses NADPH generated by glucose-6-phosphate dehydrogenase (GPDH) and 6-phosphogluconate dehydrogenase in the pentose phosphate pathway. Glutathione protects the cell membrane by rapidly destroying any peroxide in a reaction catalysed by the enzyme glutathione peroxidase (GP):

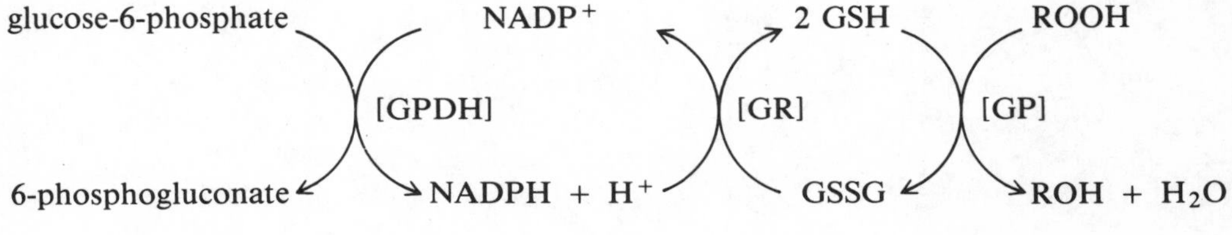

Glutathione also prevents the oxidation of the —SH groups in the membrane and cellular proteins and contributes towards keeping methaemoglobin at a low concentration in the cell.

PROTECTION OF HAEMOGLOBIN

During the oxygenation of haemoglobin a small amount becomes oxidized to methaemoglobin in which the iron is in the ferric form. This oxidized form of haemoglobin is unable to carry oxygen to the tissues so that high concentrations of methaemoglobin are dangerous and must be avoided. The amount of haemoglobin oxidized spontaneously *in vitro* is about 2 per cent per day but the concentration of the oxidized form *in vivo* is less than 1 per cent of the total haemoglobin. This low concentration is maintained by the reduction of methaemoglobin by the enzyme *methaemoglobin reductase* (MR) which uses NADH generated in glycolysis. There is a minor contribution by an NADPH-dependent methaemoglobin reductase and some reduction of methaemoglobin also takes place with glutathione and ascorbate.

$$NADH + H^+ + 2MetHb(Fe^{3+}) \xrightarrow{[MR]} NAD^+ + 2Hb(Fe^{2+})$$

ENZYME DEFICIENCY AND HAEMOLYTIC ANAEMIA

The effect of a low concentration of glutathione on the integrity of the red cell can be seen in cases of glucose-6-phosphate dehydrogenase deficiency. When this occurs there is not enough NADPH produced to keep the glutathione in the reduced state so that methaemoglobin accumulates, —SH groups of proteins are oxidized and the cell membrane is subject to lipid peroxidation. Such a concerted attack on the cell leads to haemolysis and anaemia. The enzyme defect occurs throughout the tissues but erythrocytes are particularly vulnerable since, unlike other cells, they are unable to synthesize fresh enzyme. The older erythrocytes are more susceptible to haemolysis as the rate of NADPH production falls with age.

Glucose-6-phosphate dehydrogenase deficiency is not uncommon and affects over 100 million males. There are a number of variants and the severity of the symptoms depends on the particular amino acid substitution. In some cases the haemolytic anaemia is chronic whereas in others an exogenous agent, such as the antimalarial drug *primaquine*, is needed to trigger a haemolytic episode. There are a number of drugs that can cause this condition and they are all capable of autoxidation. This results in an increase in hydrogen peroxide and other products which prove too much for an already severely depleted glutathione to cope with.

10.3 Foreign organic compounds: their metabolism and inactivation

Xenobiotic compounds

The previous two sections have considered how organisms protect themselves against damage by ammonia and oxygen which are part of normal metabolism. In this section and in Section 10.4 we shall see how animals deal with foreign compounds. *Xenobiotic compounds* (Gk *xeno* = foreign, *bios* = life) are not part of the make-up or the normal metabolism of a particular organism but that does not necessarily mean they are not found in nature. For example, antibiotics, plant alkaloids, snake venoms and natural carcinogens such as the aflatoxins are quite foreign to man and other animals but originate from the biosphere. Other xenobiotics however are synthetic compounds which are not found in nature at all and include many industrial and agricultural chemicals manufactured as drugs, food additives, pesticides and herbicides. The activity of mankind on our planet means that xenobiotic compounds, whether natural or synthetic, are deliberately or sometimes inadvertently brought into contact with the animal kingdom and many of these compounds are physiologically active or toxic. In this and the following section we shall see how these xenobiotics are metabolized and inactivated.

IMPORTANCE OF METABOLISM

The biological effect of a foreign compound, whether it is a drug or a toxin, depends on the concentration at the site of action and this in turn depends on how the compound is absorbed, distributed, metabolized and eliminated.

Xenobiotic compounds such as drugs are generally lipid soluble and poorly ionized so that they are easily absorbed from the gastrointestinal tract. These physical properties also ensure that they are readily absorbed from the glomerular filtrate produced by the kidney and so remain within the organism. During metabolism, drugs are converted to compounds that are less lipid soluble and more polar so that they can be excreted by the kidney. An example of this can be seen in the case of the anaesthetic *thiopental*, almost all of which is metabolized and less than 1 per cent of which is excreted unchanged. This is just as well since without metabolism, the biological half-life in man would be about 100 years and anaesthetized patients would be a long time regaining consciousness, if ever!

PHASE I AND PHASE II ENZYMES

There are many enzymes involved in the metabolism of xenobiotic compounds but they fall into two broad categories known as the phase I and the phase II enzymes (Fig. 10.5). The phase I enzymes introduce and/or remove a group from a foreign compound while the phase II enzymes add on a polar side chain by *conjugation*. Reactions involving phase I enzymes are considered in this section and the phase II conjugations in Section 10.4.

Mixed function oxidases

The most important reactions of the phase I enzymes are oxidations by *mixed function oxidases* in which one atom of molecular oxygen is introduced into the compound and

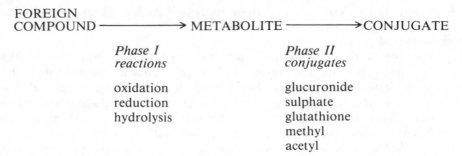

FOREIGN
COMPOUND ───────────→ METABOLITE ──────────→CONJUGATE

Phase I *reactions*	*Phase II* *conjugates*
oxidation	glucuronide
reduction	sulphate
hydrolysis	glutathione
	methyl
	acetyl

Figure 10.5   Phase I and phase II enzymes.

the other atom is reduced to water by an electron transport chain starting with NADPH (Fig. 10.6). The result of this is the hydroxylation of the substrate so the enzymes are also called *hydroxylases*. Mixed function oxidases have a very wide specificity and are responsible for the hydroxylation of steroids, drugs and a whole host of foreign compounds. Most oxidations take place on the smooth endoplasmic reticulum and a vital part of the enzyme complex is cytochrome P-450.

CYTOCHROME P-450

This is the most abundant cytochrome in the liver and is named after the intense absorption band at 450 nm seen when the cytochrome is exposed to carbon monoxide. Cytochrome P-450 is widely distributed and is found at the points of entry

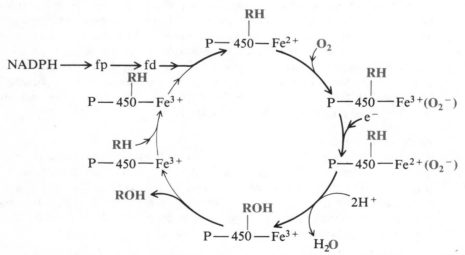

Figure 10.6   The oxidation of a foreign compound (RH) by a cytochrome P-450 mixed function oxidase (bold arrow indicates flow of electrons (e^-) from electron transport chain containing flavoprotein, fp, and ferredoxin, fd).

to the body (skin, gastrointestinal tract, lung) and the organs associated with excretion and elimination (kidney and placenta). It is also present in the adrenal glands and the gonads where it plays an important role in the hydroxylation of steroids (Section 9.4). The greatest amount however is found in the liver, the central organ of metabolism.

Cytochrome P-450 is part of an enzyme complex containing an electron transport chain which transfers electrons to the cytochrome from NADPH and to a lesser extent from NADH. The mechanism of action is still not clear but is probably similar to that shown in the diagram (Fig. 10.6).

INDUCTION OF MICROSOMAL ENZYMES

The activity of the mixed function oxidases depends on the diet and also previous exposure to xenobiotic compounds such as phenobarbitone, nicotine or DDT. For example, rats fed phenobarbitone show a proliferation of the endoplasmic reticulum in the liver and a marked increase in the cytochrome P-450, the extent of which depends on the diet (Table 10.2). This in turn determines how rapidly a foreign compound is metabolized and therefore its biological action. If a compound is inactivated by microsomal enzymes, then previous exposure to phenobarbitone will lower its biological activity. Conversely, if a compound is activated by microsomal enzymes the biological activity will be increased. The importance of the activity of the microsomal enzymes is most strikingly seen with compounds that are toxic or whose metabolites are toxic. For example, the amount of CCl_4 that will kill 50 per cent of a population of rats (LD_{50}) is normally about 1.3 ml but increases to 3 ml in starving animals and falls to a mere 0.1 ml in rats exposed to phenobarbitone. In the case of CCl_4 the LD_{50} is smaller the higher the level of cytochrome P-450 in the liver (Table 10.2) suggesting that the toxic entity is a metabolite and not the administered compound. In fact it is now thought that the liver damage is caused by the free radical CCl_3^- produced during metabolism.

Table 10.2 Factors affecting the cytochrome P-450 content of rat liver (The values shown are typical results within the range normally found experimentally)

Diet	Phenobarbitone* treatment	Cytochrome P-450 (nmol/g)
Normal	No	28
Normal	Yes	130
Starvation	No	7
Starvation	Yes	24

* 1 mg/ml in the drinking water for 1–2 weeks.

Drug metabolism

Some examples of how metabolism affects the biological activity can be seen in the following examples of drug metabolism. Only one particular reaction is considered but it should be remembered that a whole range of metabolites are normally produced *in vivo* and that their composition depends on the sex, age and species under investigation.

ACTIVATION

More than 50 years ago, it was shown that the drug *prontosil red* cured streptococcal infections in mice but had no direct effect on the bacteria *in vivo*. This was because the antibacterial compound was not the dye but *sulphanilamide*, one of its metabolites. Sulphanilamide proved to be too toxic to be used in man but chemical modification of this structure gave rise to a whole range of *sulphonamides* some of which are still in use today.

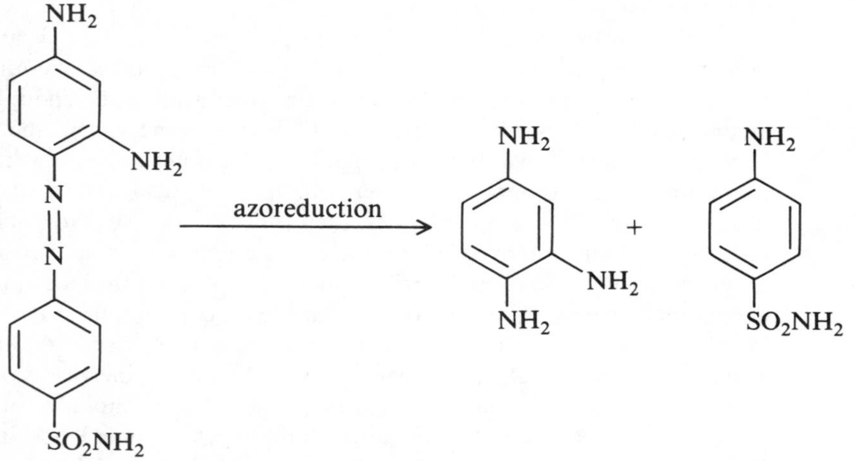

prontosil red   triaminobenzene   sulphanilamide

INACTIVATION

Most drugs are inactivated during metabolism and this is illustrated in the case of *phenobarbitone*, a powerful hypnotic. In the liver, an —OH group is introduced into the benzene ring and this makes it inactive pharmacologically:

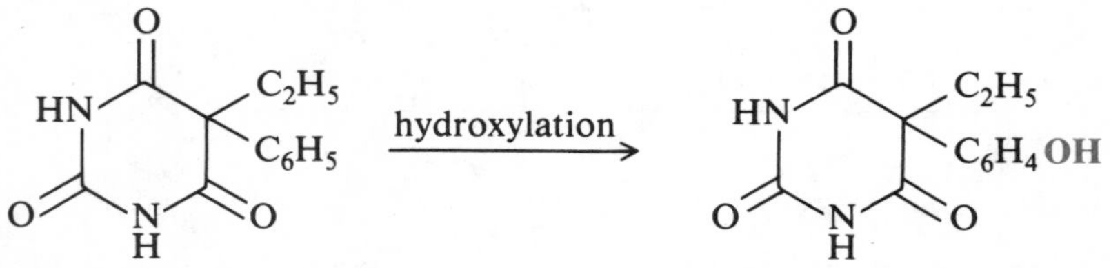

phenobarbitone   hydroxylated phenobarbitone

CHANGE OF ACTIVITY

Occasionally metabolism changes the pharmacological activity of a drug and *iproniazid*, which is an antidepressant, is dealkylated to *isoniazid* which is antitubercular:

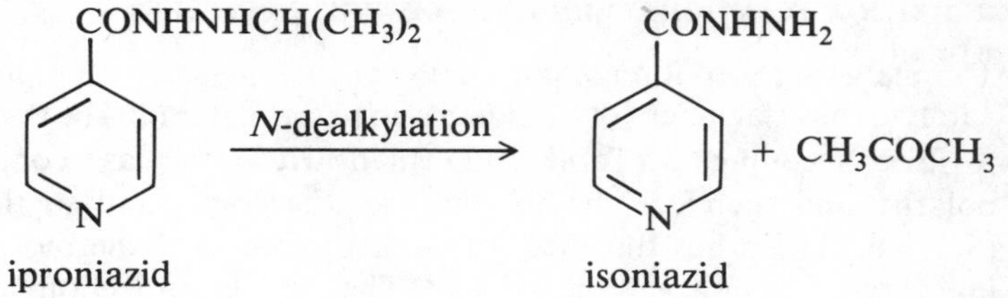

iproniazid → isoniazid + CH_3COCH_3

N-dealkylation

INTOXICATION

The analgesic and antipyretic *phenacetin* was used for many years but was eventually banned from drug formulation because of its nephrotoxicity. One of the metabolites of phenacetin is 2-hydroxy phenetidine which increases the formation of methaemoglobin. In most people the concentration of this metabolite is quite low but a few individuals are unfortunate enough to produce larger quantities than normal and this causes haemolysis and anaemia.

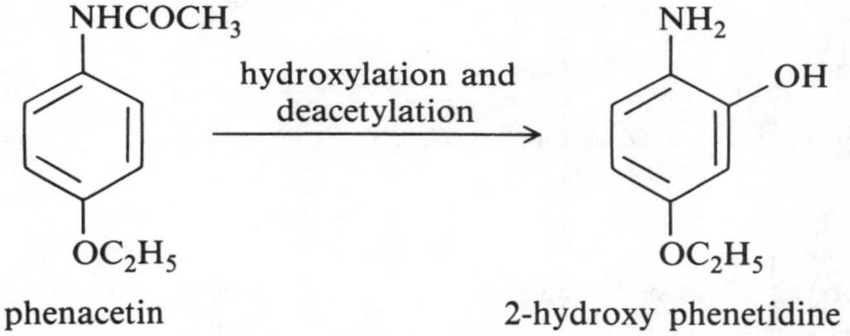

phenacetin → (hydroxylation and deacetylation) → 2-hydroxy phenetidine

Metabolism which increases the toxicity of a compound is not always a disadvantage and an example of this is seen with *malathion* which is a powerful insecticide but has a low toxicity in mammals. The reason for this is that insects convert malathion to *malaoxon* which is a powerful inhibitor of acetylcholinesterase whereas mammals metabolize malathion to the monoester and the dicarboxylic acid.

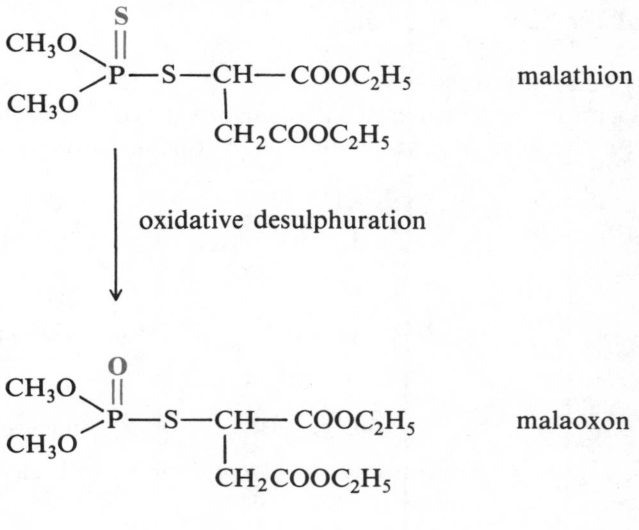

malathion

oxidative desulphuration

malaoxon

10.4   Conjugation reactions: inactivation and excretion

The metabolism of foreign compounds is completed by the phase II enzymes which catalyse the attachment of a polar group to the metabolite or the xenobiotic compound to form a conjugate (Table 10.3). In nearly every case conjugation increases the polarity and therefore the solubility of the compound so that it is more readily excreted. It also has the effect of markedly reducing the biological activity and the majority of conjugates are inactive. The synthetic reactions of conjugation are all endergonic and energy is used to 'activate' the conjugating molecule or the xenobiotic compound.

Table 10.3   Conjugation reactions catalysed by phase II enzymes

Conjugation	Phase II enzymes	Functional groups	
Glucuronide	UDP-glucuronyltransferase	—OH —NH$_2$ —SH	—COOH —NHR
Sulphate	Sulphotransferase	—OH —NHR	—NH$_2$ —SH
Glutathione	Glutathione-S-transferase	—Cl —I	—Br —NO$_2$
		—CH—CH— (epoxide)	
Methyl	Methyltransferase	—NH$_2$ —SH	—OH
Acetyl	Acetyltransferase	—NH$_2$	—OH

Types of conjugation

There are several different types of conjugation (Table 10.3) and in this section we shall see how important these reactions are in the inactivation and excretion of foreign and also natural compounds.

GLUCURONIDE SYNTHESIS

Glucuronic acid, an oxidized derivative of glucose, is a highly polar compound due to the ionized carboxyl group and the hydrophilic hydroxyls.

β-D-glucose β-D-glucuronate

These groups can form ester or glycoside links with alcohols, phenols, carboxylic acids, amines and thiols so that glucuronide formation takes place with a wide range of compounds. Examples of drugs which are excreted as their glucuronides include aspirin, morphine and chloramphenicol. Glucuronide formation is quantitatively the most important of the phase II reactions and occurs in most mammals although it is quite low in the cat. Glucuronide synthesis is catalysed by the microsomal enzyme *UDP-glucuronyltransferase* (*UDP-GT*) which uses *uridine diphosphoglucuronic acid* (*UDP-GA*) as the glucuronyl donor. UDP-GA is formed from UDP-glucose, a metabolite of glycogen synthesis, and one molecule of ATP is hydrolysed for each molecule of glucuronide produced (Fig. 10.7).

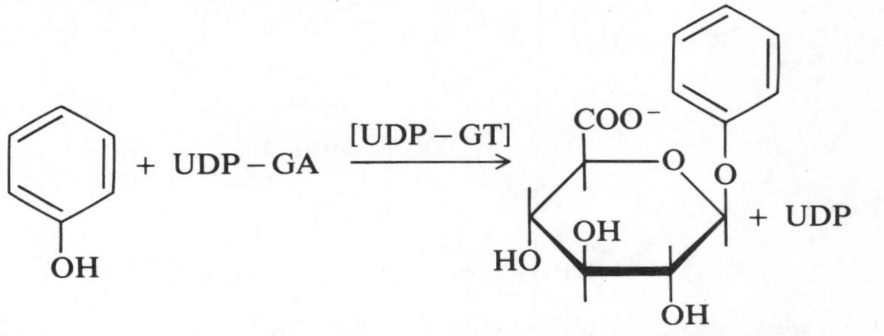

phenol

phenol-β-D-glucuronide

SULPHATE FORMATION

Sulphate conjugation is an important reaction for phenols, alcohols and amines and the resulting sulphate esters are completely ionized and very water soluble. The conjugates are formed by the action of *sulphotransferases* (*ST*) present in the cytosol which catalyse the transfer of a sulphate group from *phosphoadenosine-5'-phosphosulphate* (*PAPS*) to the compound to give the sulphate ester and phosphoadenosine-5'-phosphate (PAP).

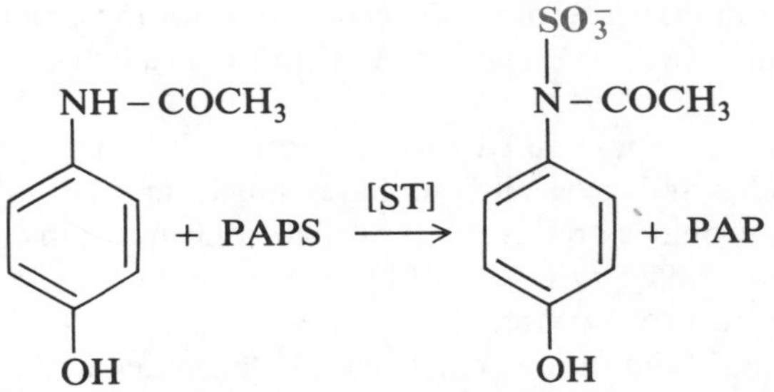

paracetamol
(drug)

paracetamol sulphate
(conjugate)

The activated molecule PAPS is synthesized from sulphate and ATP by the action of *ATP-sulphurylase* (*ATP-S*) and *adenosine-5'-sulphatophosphate kinase* (*ASP-K*):

$$SO_4^{2-} + ATP \xrightarrow{[ATP\text{-}S]} APS + PP_i$$

$$APS + ATP \xrightarrow{[ASP\text{-}K]} PAPS + ADP$$

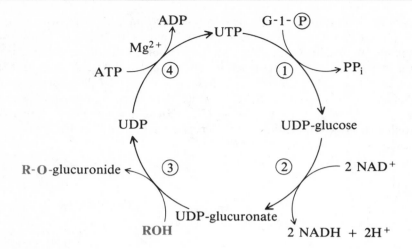

① = UDP-glucose pyrophosphorylase

② = UDP-glucose dehydrogenase

③ = UDP-glucuronyl transferase

④ . = nucleoside diphosphate kinase

UDP = uridine diphosphate

G-1-Ⓟ = glucose-1-phosphate

Figure 10.7   The formation of glucuronide conjugates.

As seen from these reactions, the energy cost for the formation of sulphate conjugates is high since two molecules of ATP are consumed for each molecule of sulphate produced.

Many xenobiotics and natural compounds have groups that can form sulphates or glucuronides and generally sulphates are formed at low substrate concentrations while glucuronides are favoured at high concentrations of substrate.

GLUTATHIONE CONJUGATION

The tripeptide glutathione reacts with halogenated and nitro compounds to form an amino acid conjugate linked through the sulphur atom of the cysteine residue (see below). The reaction is catalysed by *glutathione-S-transferase* (*GST*) and the product can then be excreted in the urine. In most cases, further metabolism takes place in which glutamate and glycine are removed by peptidases and the amino group of the residual cysteine is acetylated to form a *mercapturic acid* which is eliminated in the bile. Xenobiotics that are dealt with in this way include methyl iodide, benzyl alcohol and the esters of maleic acid.

$$\text{Glu} - \text{Cys} - \text{Gly}$$

2,4-dinitro-1-chlorobenzene + GSH $\xrightarrow{\text{[GST]}}$ glutathione conjugate + H^+ + Cl^-

2,4-dinitro-1-chlorobenzene   glutathione conjugate

METHYLATION AND ACETYLATION

These conjugates are unusual in that the products are less polar than the original compound. The other common feature is that they are quite important for endogenous compounds but are relatively minor pathways for xenobiotics. Methylation is a widespread reaction which takes place in a number of organs and is important in the metabolism of many natural compounds.

Methylation is not so common with xenobiotics but it does occur and the methylation of toxic elements is an important mechanism in their biotransformation. The non-metals arsenic and selenium undergo methylation and also a number of metals including lead, tin and mercury. An example of this is the methylation of mercury which is catalysed by enzymes present in bacteria found in the mud of rivers and lakes:

$$Hg^{2+} \longrightarrow CH_3Hg^+ \longrightarrow (CH_3)_2Hg$$

| inorganic | monomethyl | dimethyl |
| mercury | mercury | mercury |

This transformation increases the toxicity of the mercury quite markedly so that the discharge of mercury from factories into the environment represents quite a serious hazard. The cofactors needed for the methylation of mercury and the other elements are derivatives of vitamin B_{12} but most organic compounds use S-*adenosylmethionine* as the methyl donor. This is formed from ATP and L-methionine by the action of *methionine adenosyl transferase* (MAT):

$$\text{ATP} + \text{L-methionine} \xrightarrow{\text{[MAT]}} S\text{-adenosylmethionine} + P_i + PP_i$$

Acetylation takes place mainly in the liver and is an important conjugation pathway for amino compounds. The reaction is catalysed by *acetyltransferases* (AT) and uses acetyl-CoA as the acetyl donor.

The acetylated derivatives are often less soluble than the parent compounds and this can cause problems. For example, the acetylated form of the antibacterial compound sulphathiazole is considerably less soluble than the original drug and crystals of the metabolite can be formed in the renal tubules causing kidney damage unless large volumes of water are drunk.

sulphathiazole   acetyl sulphathiazole

The metabolism of bilirubin a natural metabolite

Conjugation reactions are also important for natural compounds and their metabolites as well as xenobiotics and many physiologically active compounds undergo
conjugation. For example, *steroids* form glucuronides and sulphates which are then
excreted in the urine while the *thyroid hormones* are also inactivated by conjugation
with glucuronic acid and sulphate. Other examples include the *biogenic amines*
adrenaline, serotonin and tryptamine, which undergo methylation, and histamine,
which can be acetylated as well as methylated. In this section we shall concentrate on
bilirubin, an important breakdown product of haemoglobin, as an example of the
importance of conjugation in the metabolism of endogenous compounds.

THE METABOLISM OF BILIRUBIN

Erythrocytes are continually being synthesized and destroyed and at the end of their
life span they are removed from the circulation by the reticuloendothelial system and
their constituents broken down. A major part of the red cell is haemoglobin and
between 6 and 7 g are degraded in man each day. The globin and the iron are removed
from the haemoglobin and the haem undergoes a series of oxidations and reductions
during which the tetrapyrrole ring is opened up to give *bilirubin*. This is transported,
bound to plasma albumin from the reticuloendothelial system, to the liver where it is
conjugated to bilirubin diglucuronide before being excreted in the bile. Further
metabolism takes place in the small intestine where faecal bacteria remove the
glucuronide and reduce the bilirubin to a series of related compounds known
collectively as *urobilinogens*. These are then oxidized to the *urobilins* and some of these
are responsible for the yellow colour of urine and the characteristic pigmentation of
faeces. A large number of bile pigments are formed and the scheme shown in Fig. 10.8
has been simplified for the sake of clarity.

CONJUGATED BILIRUBIN

Bilirubin is quite an insoluble molecule but esterification of the two propionic acid
residues (Fig. 10.9) by glucuronyl transferase gives bilirubin diglucuronide which is
water soluble. This is an important metabolic reaction since it enables bilirubin to be

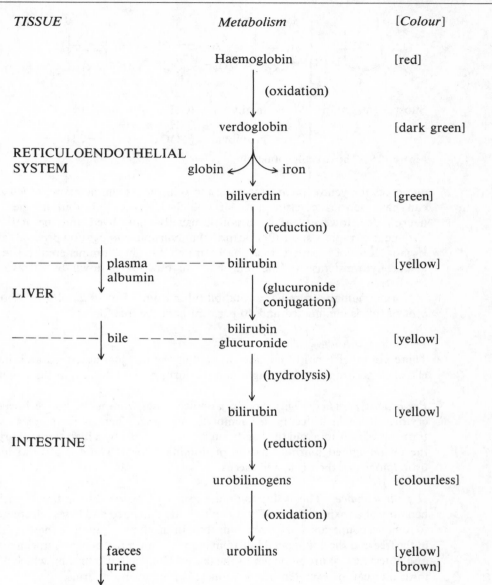

TISSUE	Metabolism	[Colour]
	Haemoglobin	[red]
	(oxidation)	
	verdoglobin	[dark green]
RETICULOENDOTHELIAL SYSTEM	globin ← → iron	
	biliverdin	[green]
	(reduction)	
plasma albumin - - - - - - -	bilirubin	[yellow]
LIVER	(glucuronide conjugation)	
bile - - - - - -	bilirubin glucuronide	[yellow]
	(hydrolysis)	
	bilirubin	[yellow]
INTESTINE	(reduction)	
	urobilinogens	[colourless]
	(oxidation)	
faeces	urobilins	[yellow]
urine		[brown]

Figure 10.8   An outline of the metabolism of the bile pigments.

excreted in a soluble form in the bile. In man the activity of glucuronyl transferase is very low or absent at birth so that the new-born infant may not be able to synthesize bilirubin diglucuronide. This means that bilirubin cannot be excreted in the bile and the circulating level rises in the blood plasma. Bilirubin then passes into the tissues

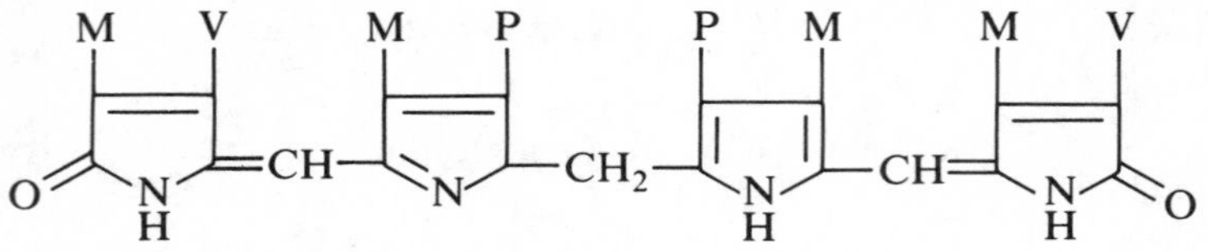

Substituent groups: M = methyl (CH_3-)
                     V = vinyl $(CH_2=CH-)$
                     P = propionate $(^-OOC-CH_2-CH_2-)$

Figure 10.9 The structure of bilirubin.

and gives the yellow pigmentation of the skin that is characteristic of *jaundice*. This can have serious consequences since high levels of bilirubin in the brain are neurotoxic. However, the condition is usually short lived and the activity of the enzyme rises rapidly soon after birth so that bilirubin conjugation proceeds as normal. If jaundice does persist then barbiturates can be administered to induce the glucuronyltransferase or the bilirubin can be broken down by ultraviolet (UV) irradiation.

Normal human serum has a total bilirubin from 3.5 to 19 μmol/litre. About 60 per cent of this is conjugated and 40 per cent is in the free form.

TYPES OF JAUNDICE

There are other types of jaundice as well as the one just described and they can be classified according to the origin and the form of the bilirubin in the circulation.

Pre-hepatic jaundice   This type of jaundice arises from an excessive breakdown of erythrocytes which occurs in haemolytic anaemia. There is therefore an increased turnover of bilirubin and the condition is characterized by a high serum bilirubin with the unconjugated bilirubin as the predominant form. There is also an increase in urobilinogen in the urine and faeces.

Hepatic jaundice   This is due to diffuse liver injury caused by a disease such as viral hepatitis or a toxin such as CCl_4. Liver cells are damaged and have a decreased ability to remove, conjugate and excrete bilirubin. In this type of jaundice there is an increase in the free and the conjugated bilirubin in the serum, a rise in urobilinogen in the urine and a decrease of urobilinogen in the faeces. Conjugated bilirubin, which is the only form that can be excreted by the kidney, is present in the urine.

Post-hepatic jaundice   This form of jaundice is caused by a blockage of bile transport due to a stone or a tumour. Bilirubin diglucuronide is unable to pass into the intestine and is absorbed into the circulation to give a rise in conjugated bilirubin in the serum. Urobilinogen is absent from the urine and also the faeces which gives them a characteristic chalky-grey appearance. There is also a marked increase in the bilirubin in the urine.

11. Molecular biology

11.1 Cell division: the replication of DNA

DNA and molecular genetics

THE TRANSFER OF GENETIC INFORMATION

One of the characteristic features of living organisms is their ability to grow, develop and reproduce. Individual cells increase in size, mass and complexity until a point is reached at which they divide. This cell division produces two identical daughter cells which are exact copies of the parent cell. Genetic information is therefore duplicated and transferred from one cell to another during cell division and for many years the nature of this hereditary material was a matter of some debate.

Experiments carried out in the last century showed that, in the case of eukaryotes, genetic information resided in the nucleus since the transplantation of a nucleus from one cell to another caused the recipient to develop the characteristic features of the donor cell. Prokaryotes do not contain a nucleus but experiments by Griffiths in 1928 demonstrated that hereditary material could be transferred from one strain of bacteria to another. He showed that a non-capsulated form of pneumococcus which was non-virulent could be converted to a capsulated virulent form by injecting mice with a mixture of live non-virulent pneumococci and dead virulent pneumococci. The injection of the live non-virulent pneumococci or the dead virulent pneumococci had no effect on the mice but the mixture proved to be lethal and the organisms isolated from the dead mice all had a polysaccharide capsule. This work was later extended by Avery and McLeod in 1949 who purified DNA from the virulent pneumococci and showed that this was the molecule responsible for the effects described.

Furthermore, the ability of different wavelengths of UV light to induce mutations coincided exactly with the UV absorption spectrum of DNA with a maximum at 260 nm. This and many other observations have since reinforced the idea that the hereditary material is DNA and we now know that the transfer of genetic information depends on the unique structure of this macromolecule.

GENES AND CHROMOSOMES

The DNA in eukaryotes is organized into discrete bodies called *chromosomes* and normally these long thin structures are spread throughout the nucleus. During cell division they become much shorter and are then visible under the light microscope as densely staining structures with a granular appearance. Human cells contain 46 chromosomes apart from the sperm and ovum which contain half this number. They are made up of *chromatin* which is a complex of DNA and protein with a small amount of RNA. Many of the proteins associated with the DNA are *histones* and there are five main species all of which have low molecular weights (11 000–21 000). Histones are particularly rich in the basic amino acids lysine and arginine and this gives them a positive charge so that they bind strongly to the negatively charged DNA.

In contrast to this, the DNA in prokaryotes is not combined with protein and is present as a single chromosome. In the case of *Escherichia coli* the chromosome is a circular double helix which is tightly packed to fit into the cell.

Each chromosome contains a large number of genes which, in classical biology, refers to the hypothetical units of hereditary material that determine the development of a particular character. Today, the term is used to describe a segment of the DNA which codes for a single polypeptide chain or a molecule of RNA. The total complement of genes in a cell is known as the *genome* and this is the same for all cells present in an organism apart from the germ cells which contain half as much. The reason why cells are structurally and functionally different is the differential expression of this common genetic information.

THE CELL CYCLE
During cell division each daughter cell receives an exact copy of the genetic information of the parent cell by replication of the DNA. In eukaryotes, the series of events leading up to and including cell division is known as the *cell cycle* and this is shown in Fig. 11.1. The cell cycle consists essentially of a *synthetic phase* (S) and *mitosis*

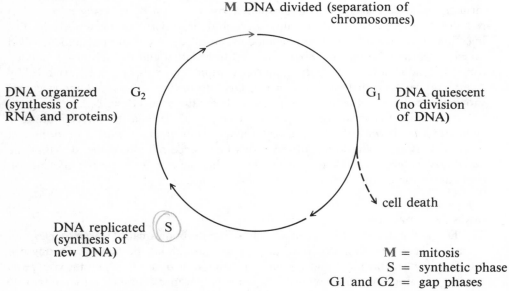

Figure 11.1 The cell cycle.

(M) separated by two *gap phases* (G_1 and G_2). The non-reproductive phase (G_1 is followed by the synthetic phase during which the amount of DNA doubles as new material is synthesized. This is followed by another gap phase (G_2) when the chromatin becomes organized into clearly visible structures prior to cell division. Mitosis (M) is visually the most dramatic part of the cell cycle and involves a series of changes in appearance of the nucleus which can be seen under the light microscope.

During mitosis the chromosomes shorten and thicken and divide evenly between the two daughter cells which then enter (G_1), the non-reproductive phase, to complete the cell cycle (Fig. 11.1). The length of G_1 depends on the particular cell but it is usually the longest part of the cycle. If G_1 is prolonged then the cells can be considered to be in a quiescent state (G_0). Virtually all eukaryotic cells go through this cycle, the length of which varies from hours to more than 100 days depending on the particular cell. Nerve cells are exceptional in that there is no further cell division once they are formed. Human and other mammalian erythrocytes do not fit into this pattern either since they lose their nucleus and therefore their ability to divide on maturation.

Mechanism of DNA replication

During cell division there is a complete duplication of the genome and the basis of this lies in the replication of the DNA. This is a complex process and although the essential features are known, there are many details still not fully understood.

MESELSON AND STAHL EXPERIMENT

Watson and Crick in 1953 showed that DNA is a double helix with the two strands held together by hydrogen bonding between complementary base pairs (Section 4.5). This means that the base sequence of one strand specifies the sequence in the complementary strand so that two identical copies of the parent DNA can be made if each strand acts as a template on which the new strands are synthesized. Evidence for this *semi-conservative* mechanism of DNA replication came from the elegant experiments of Meselson and Stahl in 1957. These workers grew *E. coli* in a medium in which the nitrogen source (NH_4Cl) was labelled with ^{15}N, the heavy isotope of nitrogen, so that after several generations the DNA contained only heavy nitrogen. The bacteria were then transferred to a medium containing ^{14}N, the normal isotope of nitrogen, and the cells harvested after each generation. The DNA was then isolated and analysed by centrifugation on a gradient of caesium chloride (CsCl). DNA containing only ^{15}N is called 'heavy DNA' while the DNA with ^{14}N is referred to as 'light DNA'. The results of this experiment showed that after one generation there was no 'heavy' or 'light' DNA but only a 'hybrid' form. After two generations there was an equal mixture of the 'hybrid' and the 'light' DNA and this pattern persisted in subsequent generations with the proportion of the hybrid form gradually diminishing. These results supported the *semi-conservative* mechanism of DNA replication (Fig. 11.2).

PREPARATION OF THE DNA TEMPLATE

From the above experiment it is clear that some form of strand separation takes place

Generation   Replication of DNA   DNA present

Original   Heavy

First   Hybrid

Second   Hybrid
                                                                        and
                                                                        light

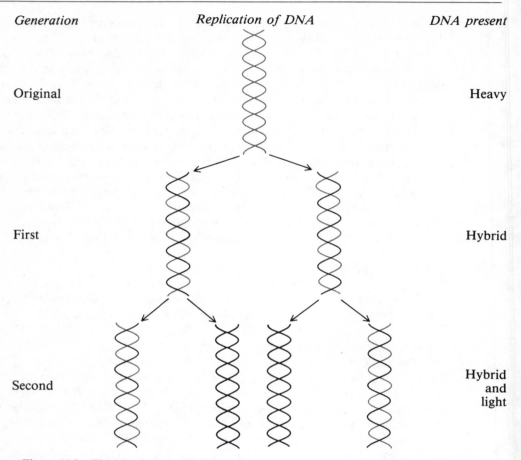

Figure 11.2   The Meselson and Stahl experiment showing that the replication of DNA is by a semi-conservative mechanism. *Escherichia coli* were grown on a medium with the heavy isotope ^{15}N (red) then transferred to a medium containing the normal isotope ^{14}N (black).

in order to copy the DNA. However, for the strands to separate completely would require a considerable expenditure of energy and it is now thought only part of the DNA unfolds to form a *replication fork*. The double helix is rapidly unwound by the enzyme *DNA-helicase* and the resulting single strands are prevented from recombining by *helix-destabilizing proteins* which bind strongly to single-stranded DNA. The DNA molecule is extremely long and the synthesis of new DNA takes place at several replication forks at once.

DNA POLYMERASE

The synthesis of new DNA on the templates is catalysed by the enzyme *DNA polymerase* (DNA-P). The substrates for this enzyme are deoxynucleotide triphosphates (dNTP) which are hydrolysed and the energy released used to synthesize the DNA:

$$n\,\text{dNTP} \xrightarrow[\text{[Mg}^{2+}]]{\text{[DNA-P]}} \text{DNA} + n\,\text{PP}_i$$

There are several enzymes which can catalyse this reaction and *DNA polymerase I* in bacteria, which was the first to be discovered, was later shown to be concerned with the repair of DNA rather than with the synthesis of new molecules. The replicase enzyme responsible for the production of new DNA in bacteria is *DNA polymerase III*. This only works from the 5' to the 3' end of the DNA template and not in the reverse direction, yet new molecules of DNA are synthesized on both strands of the parent DNA. The DNA on the *leading strand*, which runs from 5' to 3', is synthesized continuously whereas the DNA on the *lagging strand*, which runs in the opposite direction, is copied in a discontinuous fashion in a series of short bursts. The pieces of the DNA produced on the lagging strand are called *Okazaki fragments* after their discoverer.

ROLE OF RNA

The DNA polymerase needs a short stretch of polymer on which to build and this is provided by an *RNA primer* synthesized by RNA polymerase. The RNA primer is later removed by ribonuclease and the gaps filled in with DNA polymerase I. The fragments are then sealed with *DNA ligase* and this is accompanied by the hydrolysis of ATP.

COMPLETION OF THE REPLICATION

The overall effect of these enzymes is to ensure that both strands of DNA are copied and that their growth proceeds together in the same direction as the moving fork. A schematic representation of DNA replication is shown in Fig. 11.3.

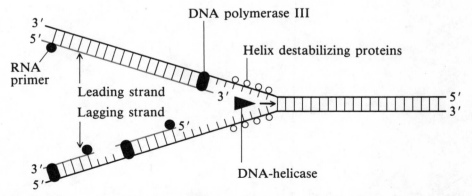

Figure 11.3 A schematic representation showing the essential features of DNA replication. The original parent DNA (black) is shown with a replication fork, and the synthesis of new DNA (red) taking place on the separated strands.

The bases in the new DNA may then be modified, for example by methylation of cytosine. The advantage of this is that any non-methylated foreign DNA entering the cell (e.g. from a bacterial virus) will be recognized as such and degraded by enzymes in the host cell. These hydrolytic enzymes are known as *restriction endonucleases* or more simply as *restriction enzymes*. Finally, the DNA is folded and packaged as chromosomes directly in prokaryotes or after binding to histones, other proteins and RNA to form chromatin in the case of eukaryotes.

11.2 Protein synthesis: transcription

Gene expression

THE FLOW OF GENETIC INFORMATION

Gene expression is the synthesis of proteins during which genetic information is transferred from the DNA to the site of protein synthesis in the cytoplasm. This involves *transcription*, during which the base sequence of the DNA is copied on to RNA, and *translation*, when the four-letter code of RNA is converted to the 20-letter code of the amino acids needed for protein synthesis. The flow of genetic information is therefore as shown in Fig. 11.4. This is quite a complex process and involves the orderly interaction of three forms of RNA, 20 amino acids, the input of energy and a whole host of enzymes and cofactors. Therefore, the account of transcription in this chapter and translation in the following chapter has been simplified in order to highlight the essential features of a complicated series of events.

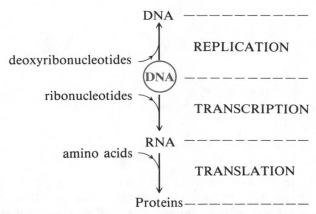

Figure 11.4 The flow of genetic information from DNA.

PROKARYOTIC AND EUKARYOTIC GENES

During transcription, the genetic information in the nuclear DNA is copied by base pairing to give RNA. The basic process is similar in all cells but eukaryotes show some additional features due to the structure of their genomes and the way they are copied. In prokaryotes, each gene is a continuous region of the DNA whereas, in eukaryotes, many structural genes are not continuous but are interrupted by non-coding regions of the DNA (Fig. 11.5). These eukaryotic genes are therefore not collinear and the regions containing the genetic information are known as *exons*, while the non-coding segments are called *introns*. Furthermore, in eukaryotes a number of genes are repeated whereas in prokaryotes there is only a single copy. The association of eukaryotic genes with chromatin and the presence of membranes in eukaryotic cells also tends to complicate the picture.

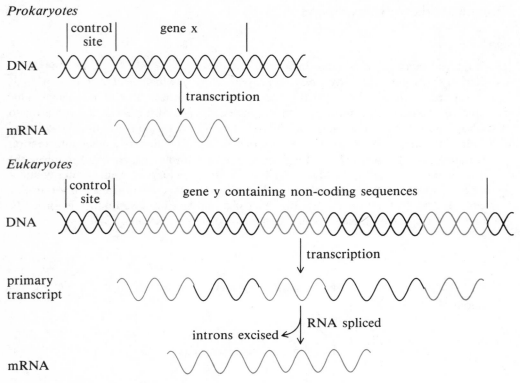

Prokaryotes

Eukaryotes

Figure 11.5 Transcription in prokaryotes and eukaryotes.

In spite of these differences there are many features of protein synthesis common to both prokaryotes and eukaryotes. All cells, for example, make three types of RNA by transcription: *ribosomal RNA* (rRNA), *transfer RNA* (tRNA), and *messenger RNA* (mRNA) and each of these three forms of RNA plays a part in protein synthesis. In *Escherichia coli*, 80 per cent of the RNA is present as rRNA, 15 per cent as tRNA and 5 per cent as mRNA.

Transcription

THE SYNTHESIS OF mRNA

The first step in the synthesis of proteins is the transfer of the genetic message from DNA to messenger RNA. In prokaryotes, this is accomplished by the synthesis of an RNA molecule with a base sequence complementary to the parent DNA, a process which uses the nucleotide triphosphates (ATP, CTP, GTP, UTP) as the building blocks for RNA, and the enzyme *RNA polymerase* (RNA-P).

If NTP represents the nucleotide triphosphates then the reaction catalysed by RNA polymerase is:

$$n \, \text{NTP} \xrightarrow[\text{[Mg}^{2+}\text{]}]{\text{[RNA-P]}} \text{RNA} + n \, \text{PP}_i$$

Only one strand of the DNA is copied and the double-stranded DNA template is preserved so that transcription is *asymmetric* and also *conservative*.

Initiation　The RNA polymerase is similar to DNA polymerase in that it operates from the 5′ to the 3′ end of the nucleic acid. The start of transcription is signalled by a sequence of base pairs that are rich in adenine and thymine and thus A-T are held together by only two hydrogen bonds as opposed to the three hydrogen bonds of each G–C pair. The bonds holding the DNA duplex at this point are therefore weaker than elsewhere which allows for the strands to separate and form a 'bubble'. This enables the RNA polymerase to identify the start of the gene and to select the strand of the DNA to be copied, which is known as the *sense strand*.

Termination　The RNA polymerase then moves along the gene incorporating one base at a time into the RNA until the termination point is reached, where another 'bubble' in the DNA formed by a particular sequence of bases signals the end of the message.

The synthesis of mRNA is similar in eukaryotes but the first RNA transcript of the DNA, known as the *primary transcript*, contains all the introns as well as the exons. The non-coding introns are then excised and the RNA sections copies from the exons are spliced together to give mRNA (Fig. 11.5).

CONTROL OF TRANSCRIPTION

There are several points at which protein synthesis can be controlled but the regulation of transcription in prokaryotes is probably the best understood. In *E. coli*, which is a typical prokaryote, the cellular DNA is present as a single circular chromosome. This contains the *structural genes* for the production of mRNA and *control genes* so that the structural genes are only expressed when the control genes allow transcription to occur. The mechanism for this is known as the *operon model* and was first proposed by Jacob and Monod in 1961.

One particular control system which has been investigated in some detail is the *lac operon* of *E. coli* and this is illustrated in Fig. 11.6. The lac operon is the cluster of genes responsible for the metabolism of lactose. It consists of three structural genes (*z, y, a*) and a control region that contains a *promoter* (p) and an *operator* site (o). RNA polymerase binds to the promoter site and immediately adjacent to this is the operator site which is the binding site for the *repressor* coded for by the regulatory gene (*i*). In the absence of an inducer, the repressor binds to the operator site and blocks RNA polymerase so that the structural genes cannot be expressed.

Such a control system enables bacteria to respond rapidly to a change in their environment so that when lactose is introduced into the growth medium the enzymes needed for its metabolism can be rapidly synthesized in the required amounts. The enzymes which vary in this way are known as *inducible enzymes*, in contrast to the *constitutive enzymes* which remain constant irrespective of the metabolic state of the cell.

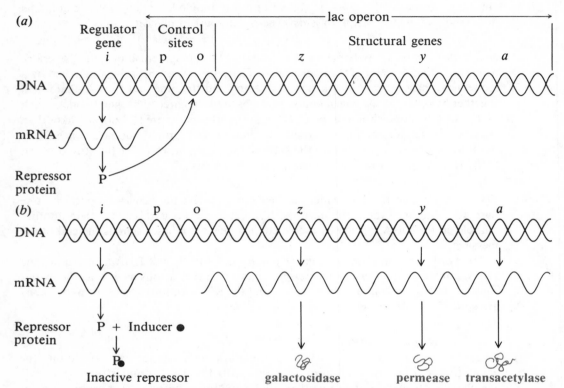

Figure 11.6   Regulation of the lac operon in *E. coli*. The diagram is not drawn to scale since the genes at the control sites are much smaller than the other genes. The direction of transcription is from left to right. (*a*) Enzyme repression (repressor binds to operator and blocks RNA polymerase). (*b*) Enzyme induction (inactive repressor means that lac operon can be transcribed and translated).

11.3   Protein synthesis: the genetic code and translation

The genetic code

A TRIPLET OF BASES

One of the problems facing early workers investigating protein synthesis was the question of translation or how RNA, made up of only four different bases, could direct the ordering of the 20 different amino acids needed for the synthesis of proteins. A simple 1:1 correspondence would only give 4 (4^1) amino acids and a 2:1 correspondence would only allow for 16 (4^2) arrangements. However, if there is a sequence of three bases coding for each amino acid, then this would produce 64 (4^3) possible arrangements which is more than enough for the 20 amino acids. Early evidence for the triplet code came from the work of Crick and his colleagues with bacterial mutants. They treated bacteria with mutagenic agents such as acridine orange and showed that if one or two bases were added or removed then no protein, or one with serious defects, was produced. However, if the mutagenic agent introduced or deleted three base pairs, then the protein was produced but with a missing or extra amino acid. This principle is illustrated below using a sentence made up of three letter words. One or two deletions produce nonsense but three deletions mean that the message can be read and makes sense.

No. of deletion
from ↓

		↓	
0	HOW THE FAT CAT DID EAT		sense
1	HOW THE ATC ATD IDE AT		nonsense
2	HOW THE TCA TDI DEA T		nonsense
3	HOW THE CAT DID EAT		sense

DECIPHERING THE CODE

The definitive work on deciphering the genetic code was largely carried out in Nirenberg's laboratory using polynucleotides with a known repeating sequence synthesized by Khorana. Early experiments showed that the addition of poly U to an *in vitro* system from *E. coli* made a protein containing only phenylalanine. Most of the code however was unravelled by the ribosome-binding technique which determined which amino acid in a radioactive mixture was bound to ribosomes and a synthetic nucleotide with a known base sequence. This enabled 61 out of the possible 64 combinations of triplet bases to be assigned to a specific amino acid. The remaining three combinations were at first called nonsense codons but are now known to signal the release of the completed polypeptide. The code which is now universally accepted is given in Table 11.1.

PROPERTIES OF THE CODE

Experimental work in a number of laboratories not only unravelled the code but also enabled a number of its important features to be identified.

Table 11.1   The genetic code showing the triplet of bases on the mRNA for each amino acid

1st Base (5′ end)	2nd Base				3rd Base (3′ end)
	U	C	A	G	
U	Phe	Ser	Tyr	Cys	U
	Phe	Ser	Tyr	Cys	C
	Leu	Ser	STOP	STOP	A
	Leu	Ser	STOP	Trp	G
C	Leu	Pro	His	Arg	U
	Leu	Pro	His	Arg	C
	Leu	Pro	Gln	Arg	A
	Leu	Pro	Gln	Arg	G
A	Ile	Thr	Asn	Ser	U
	Ile	Thr	Asn	Ser	C
	Ile	Thr	Lys	Arg	A
	Met	Thr	Lys	Arg	G
G	Val	Ala	Asp	Gly	U
	Val	Ala	Asp	Gly	C
	Val	Ala	Glu	Gly	A
	Val	Ala	Glu	Gly	G

Sequential   The code is *non-overlapping* and is read in blocks of three bases at a time.

Universal   The genetic code is *ubiquitous* and applies to proteins synthesized in bacteria, yeast, animal and plant cells. The only exception to this are some minor variations in mitochondria and chloroplasts.

Degenerate   The code is degenerate in that nearly all amino acids are coded for by more than one base triplet and in the majority of cases the first two bases appear to be the most important.

Unambiguous   The code is not ambiguous since a particular triplet of bases codes for only one amino acid.

The stages of translation
The translation of the four-base code of the nucleic acids to the 20-amino-acid code of

1. Complex formed between tRNA carrying an amino acid ①, mRNA and the ribosome.

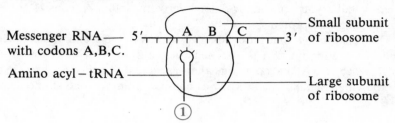

2. Transfer RNA carrying the second amino acid ②, binds at codon B.

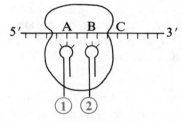

3. Peptide bond formed, free tRNA released and ribosome moves along mRNA by translocation.

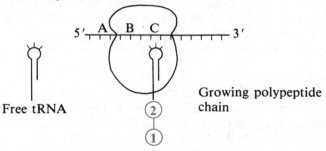

Figure 11.7 A schematic representation of protein synthesis.

proteins is the next stage of protein synthesis and takes place in a complex series of interrelated steps (Fig. 11.7).

ACTIVATION

The first step is the activation of an amino acid to form a complex with tRNA in a reaction catalysed by *aminoacyl-tRNA synthetase* (AAS):

$$\text{amino acid + tRNA} \xrightarrow[\substack{\text{ATP} \quad \text{AMP + PP}_i}]{\text{[AAS]}} \text{amino acyl−tRNA}$$

The first amino acid to be incorporated is methionine in eukaryotes and *N*-formyl methionine in prokaryotes. The methionine is later removed from proteins synthesized by eukaryotes but the *N*-formyl derivative often remains as the first amino acid in prokaryotic proteins.

All tRNAs have the same base sequence —CCA at the 3′ end of the molecule and it is here that the amino acid is attached via the 2′ or 3′ of the terminal adenosine. However, each tRNA binds only one amino acid and this specificity lies in a triplet of bases known as the *anti-codon* which is complementary to the amino acid codon of mRNA.

INITIATION

The anti-codon of the methionine–tRNA complex binds to the codon AUG of the mRNA. Immediately prior to this initiation site is the sequence AGGA which binds to the complementary sequence UCCU present in the small subunit of the ribosome. This is then followed by the binding of the large ribosomal subunit to give the complete complex. Various protein *initiation factors* are involved and energy is provided by the hydrolysis of ATP and GTP.

ELONGATION

A tRNA carrying the second amino acid now binds to the next codon towards the 3′ end of the mRNA and involves an *elongation factor* and the hydrolysis of GTP. The two amino acids now react to form a peptide bond during which the first amino acid is transferred to the second tRNA and the first tRNA is released. This reaction is catalysed by the enzyme *peptidyl transferase* which is one of the proteins of the large ribosomal subunit. Another elongation factor is bound, GTP is hydrolysed and the ribosome moves three bases along the mRNA by *translocation* to reveal the next codon ready for the tRNA carrying the third amino acid.

TERMINATION

The whole process is repeated for each amino acid until one of the termination codons is reached. The completed polypeptide is then released from the last tRNA, GTP is hydrolysed and the ribosomal subunits dissociate from the mRNA. The various stages of protein synthesis illustrated in the diagram (Fig. 11.7) show the sequence of events at a single ribosome but in practice the mRNA is often translated by several ribosomes simultaneously. This increases the rate of synthesis of the polypeptide and such a group of ribosomes strung out like beads along the thread of mRNA are known as *polysomes* (Fig. 11.8).

ENERGY REQUIREMENTS

The synthesis of proteins is very expensive in terms of the energy expended. Just how costly can be seen in Table 11.2 which shows how much ATP and GTP is needed for the incorporation of just one amino acid into the macromolecule. The real cost in terms of energy is actually much higher than this, since metabolic energy is needed to synthesize the various components of the machinery of translation and this has not been taken into account.

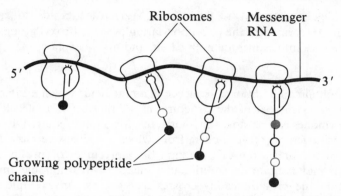

Figure 11.8 The synthesis of several copies of a polypeptide on one polysome.

Table 11.2 The energy requirements of translation

Stage of translation	Molecule hydrolysed nucleotide	No.
Activation	ATP	2*
Initiation	ATP	1
	GTP	1
Elongation	GTP	2
Termination	GTP	1

Equivalents of ATP for each peptide bond formed = 7

* ATP → AMP + PP$_i$ equivalent to two molecules of ATP → ADP + P$_i$.

Post-translational changes

The polypeptide released at this stage may have the primary structure of the protein present in the cell but in many cases the molecule undergoes some modification before being used by the organism.

MODIFICATION OF AMINO ACIDS

Proline is synthesized in the *trans* configuration and then undergoes slow isomerization to the *cis* form. Other proline residues such as those in collagen are hydroxylated to hydroxyproline. Some histidine residues may be methylated and the thiol groups of two cysteine residues close to each other may be oxidized to form a strong covalent disulphide bond.

FORMATION OF CONJUGATED PROTEINS

Glycoproteins, such as those found in mucin, are formed by the addition of oligosaccharide side chains through *O*-glycoside linkages to serine and threonine residues. The plasma glycoproteins are another major class of conjugated proteins and here the oligosaccharides are linked via *N*-glycoside bonds to asparagine. Phosphoproteins are formed by the phosphorylation of the hydroxyl groups of serine and threonine and examples include milk casein and egg phosphovitin.

HYDROLYSIS

Some proteins contain an additional sequence of amino acids which is later removed by hydrolysis. The extra part of the structure means that active molecules such as enzymes and hormones can be stored in an inactive form until required. Examples of this include pepsinogen and trypsinogen, from which the digestive enzymes pepsin and trypsin are made, and proinsulin the immediate precursor of insulin.

Some proteins may contain an additional sequence of hydrophobic amino acids which means that the protein readily crosses membranes. An example of this is preproinsulin which contains 23 more amino acids than proinsulin.

FOLDING OF THE POLYPEPTIDE CHAIN

The final stage of protein synthesis is the folding of the polypeptide chain to form the secondary and tertiary structure of the protein. This occurs spontaneously and the final shape of the molecule is determined by the primary structure which is the order of the amino acid residues. Proteins such as haemoglobin, with a quaternary structure, are formed at this stage as the subunits come together to form the completed molecule.

11.4 Inhibitors of nucleic acid metabolism: chemotherapy

There are a number of compounds that inhibit DNA replication and protein synthesis and they are invariably highly toxic to the affected organism (Table 11.3). Some of these inhibitors however have been successfully used in the treatment of disease by *chemotherapy*. These drugs act by exploiting subtle differences in the nucleic acid metabolism of different cells so that the growth of invading organisms or tumours is inhibited while that of the normal cells of the host remain unaffected. However, this is an ideal that is not always realized since many drugs have adverse side effects and a small number of individuals may react badly to a particular compound. Fortunately, these side effects are relatively minor for most antibiotics but are still a major problem for anti-cancer agents. The search therefore continues for the ideal chemotherapeutic agent for each disease: one that has a high degree of selective toxicity with no side effects.

Table 11.3 The inhibitory action of antibiotics and anti-metabolites

Binding site	Inhibitor
Cellular DNA	
Intercalation	Actinomycin D
Intercalation and distortion	Chloroquine
Covalent binding and cross-linking	Nitrogen mustards
Enzymes	
DNA polymerase	Nalidixic acid
RNA polymerase	Rifampicin
Ribosomes	
30 S subunit	Streptomycin
	Tetracyclines
50 S subunit	Chloramphenicol
	Erythromycin
60 S subunit	Cycloheximide

Cancer chemotherapy

CANCER

The ability of DNA to replicate accurately without errors is obviously a vital part of the life of the cell and any interference with this process will have potentially disastrous consequences. Either the cell will die, or worse still, it may continue to divide without restraint and develop into a tumour. The molecular mechanisms responsible for the development of cancer are still not fully understood but undoubtedly involve the DNA since compounds that are carcinogenic in man either damage DNA directly or interfere with its replication. Alterations to the gene, known as *mutations*, occur spontaneously with a frequency of about one in a million for each cell division and this rate can be markedly increased by carcinogens or atomic and UV radiation. Fortunately, tumours are not as common as mutations since the damaged gene is normally repaired by the excision of the faulty region of the DNA and its replacement with normal bases. Furthermore, multiple mutations are needed not just one, before a cell can become cancerous. However, tumours do develop and cancer is responsible for about 20 per cent of deaths in developed countries. Cancer research has therefore been a high priority in medical laboratories and a number of drugs are now available to treat this condition. Many of these anti-cancer agents act by blocking the replication of DNA in the rapidly dividing tumour cells but unfortunately these powerful drugs are also toxic and interfere with the DNA of normal healthy cells. Their dose therefore has to be carefully controlled so that it is high enough to kill the tumour but low enough to minimize the damage to healthy tissue.

ANTI-METABOLITES

These drugs are structural analogues of normal metabolites and interfere with the supply of raw materials needed for DNA synthesis. One example of this type of anti-cancer drug is *5-fluorouracil* used to treat skin cancer. This is metabolized to 5-fluorodeoxyuridine monophosphate which inhibits thymidylate synthase (TS) and blocks the methylation of deoxyuridine monophosphate (dUMP) to form deoxythymidine monophosphate (dTMP). Cells deprived of this nucleotide, which is one of the building blocks of DNA, are unable to replicate and eventually die.

$$\text{dUMP} \xrightarrow[\substack{\text{folic acid} \\ \text{derivatives}}]{\text{[TS]}} \text{dTMP} \dashrightarrow \text{DNA}$$

Another example of an anti-metabolite is *cytosine arabinoside* (cytarabine) used in the treatment of myelocytic leukaemia. This inhibits the formation of deoxycytosine monophosphate (dCMP) and therefore blocks the synthesis of DNA.

5-fluorouracil

cytosine arabinoside

COMPOUNDS THAT REACT WITH DNA

These drugs interfere with cell growth by reacting directly with the DNA. Some do so by *intercalation* in which their flat aromatic rings are inserted between base pairs and bind strongly to the DNA. The classical example of this type of compound is *actinomycin D* whose phenoxazone ring intercalates at the G–C pairs of DNA and blocks the synthesis of RNA. The compound is therefore very toxic but it has been used to treat Wilm's tumour.

X = cyclic peptide
- L-methylvaline
- sarcosine
- L-proline
- D-valine
- L-threonine

Other compounds, such as the *nitrogen mustards*, react with the DNA by forming cross-links between the guanine bases which completely block both replication and transcription. They were first developed as war gases but have now found a more peaceful and constructive application as anti-cancer agents and are particularly useful in the treatment of Hodgkin's disease.

1

a nitrogen mustard

There are of course many other drugs used in the chemotherapy of cancer and this area of biochemical pharmacology will continue to be explored as more is discovered about the molecular mechanisms of cell division and its control.

Antibiotics

The term antibiotic literally means against life (Gk *anti* = against, *bios* = life) and as the name implies antibiotics are quite toxic compounds. This toxicity however is highly selective so that invading bacteria are killed while the cells and tissues of the host are unaffected. Antibiotics interfere with bacterial metabolism in a variety of ways but those considered below are all inhibitors of protein synthesis.

INHIBITORS OF TRANSCRIPTION

Antibiotics that interfere with transcription do so either by binding to the DNA and preventing an RNA copy being made or by inhibiting RNA polymerase directly.

Adriamycin is an example of a compound that blocks the synthesis of RNA by intercalating the DNA. It is used clinically to treat solid tumours but suffers from the disadvantage that it is toxic to the heart.

Chloroquine is an anti-malarial drug which binds strongly to the DNA of the malarial parasite by intercalation. Its effectiveness and specificity is due to the fact that it is concentrated as much as 1000 times by infected erythrocytes so that the parasite is exposed to very high concentrations of the drug.

Rifampicin also inhibits transcription but this time by binding directly to the RNA polymerase. It is active against Gram-positive bacteria and is particularly useful for dealing with infections caused by *Mycobacterium tuberculosis*.

INHIBITION OF TRANSLATION

Many antibiotics which inhibit translation bind to ribosomes and owe their selectivity

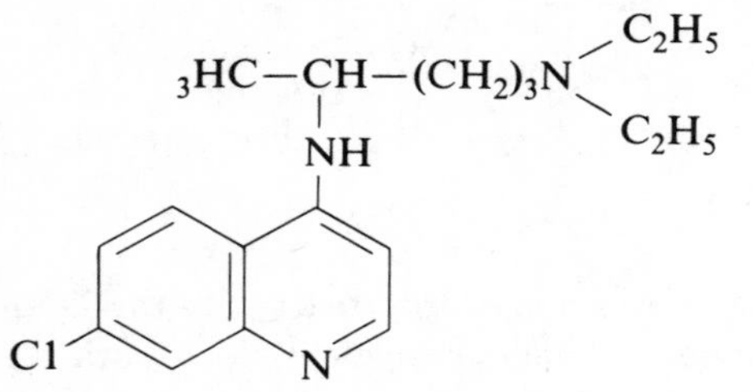

chloroquine
(antimalarial drug)

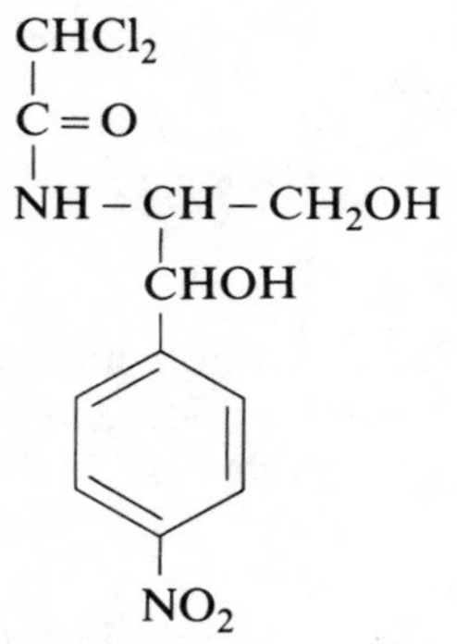

chloramphenicol
(antibacterial compound)

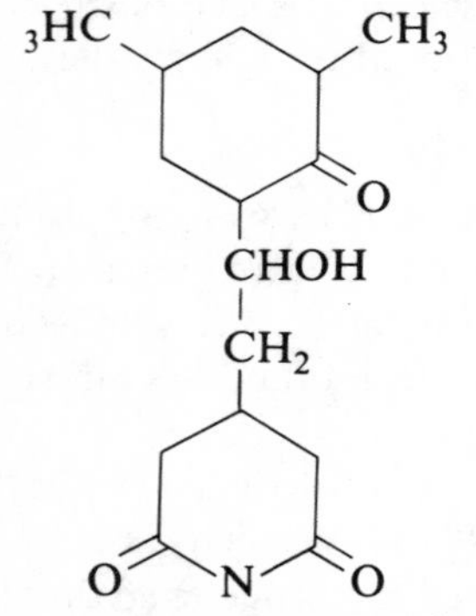

cycloheximide
(antifungal agent)

Figure 11.9 The structure of some antibiotics.

to the difference in the size and composition of ribosomes in prokaryotes and eukaryotes (Section 5.5).

Streptomycin and *erythromycin* bind strongly to the 30 S subunit of bacterial ribosomes. Streptomycin causes the message of mRNA to be misread while erythromycin inhibits translocation. The *tetracyclines* are another group of antibiotics which bind to the 30 S subunit of prokaryotic ribosomes and prevent the attachment of the amino acyl-tRNA. They also interact with the 40 S subunit of eukaryotic ribosomes *in vitro* but they have little effect on eukaryotic protein synthesis *in vivo*. The reason for this specificity is that bacteria are far more permeable than animal cells to tetracyclines.

Chloramphenicol, on the other hand, binds tightly to the larger 50 S subunit of prokaryotic ribosomes and inhibits peptidyltransferase activity. *Cycloheximide* is similar to chloramphenicol in that it also inhibits peptidyl transferase but this time on the large 60 S subunit of eukaryotes. Cycloheximide is therefore toxic to animal cells but is useful as an external anti-fungal agent.

The structures of some of the simpler antibiotics are given in Fig. 11.9.

11.5 Applied molecular biology: genetic engineering

Inhibitors interfere with DNA metabolism by blocking enzymes but a more subtle way of directing the metabolism of DNA is to manipulate the gene. This is popularly known as *genetic engineering* and basically involves the introduction of foreign DNA into a cell in order to change its genetic make-up.

Recombinant DNA

Genetic engineering is also called *recombinant DNA technology* since a foreign gene is introduced as a hybrid molecule of recombinant DNA. This hybrid is formed when a fragment of DNA from one source is covalently linked with the DNA of another organism. The synthesis and isolation of recombinant DNA has been made possible by bringing together a number of techniques of molecular biology.

RESTRICTION ENZYMES

One of the most important tools is the use of *restriction endonucleases* which cleave DNA at specific sites represented by a particular sequence of bases in the macromolecule. The nucleases are produced by, and therefore obtained from, prokaryotes and their function in the host cell is to attack and destroy foreign DNA. The enzymes do not affect the normal cellular DNA since potential cleavage points are protected by methylation. These enzymes do not hydrolyse the DNA completely but give rise to a number of 'gene sized' fragments.

The 'cuts' (↓) produced by the endonuclease may be found at the same position in each strand of the duplex to give 'blunt ended' fragments or they may be staggered to give what are known as overlapping or 'sticky ended' fragments and it is this second group of enzymes that have found most application in genetic engineering.

Endonuclease HaeI from Haemophilus aegyptus

5′ − C − G ┼ C − C − 3′   'Blunt ended' fragments
3′ − G − C ┼ G − G − 5′

Endonuclease EcoRI from Escherichia coli

5′ − G ┼ A − A − T − T − C − 3′   'Sticky ended' fragments
3′ − C − T − T − A − A ┼ G − 5′

PREPARATION OF THE GENE

The first task is to obtain the required gene. This can be prepared by cleaving genomic DNA (isolated from an appropriate tissue) with restriction endonucleases and selecting out the fragment containing the desired gene. Alternatively, the gene may be produced by purifying the mRNA and making a DNA copy. The normal flow of genetic information is from DNA to RNA (Fig. 11.7) but there are enzymes found in *retroviruses* which will catalyse the reverse process given a supply of deoxynucleotide triphosphates (dNTP). The enzymes catalyse the reverse of normal transcription and are therefore known as *reverse transcriptases* (RT). The DNA copy (cDNA) made from the RNA is only single stranded so this is incubated with DNA polymerase (DNA-P) and deoxyribonucleotide triphosphates (dNTP) to produce the double-stranded DNA corresponding to the required gene. This method however provides a gene which is shorter than the normal genomic gene. It is shorter because the mRNA from which it was derived lacks the intron sequences and control regions that would be associated with the genomic copy of the gene.

$$\text{mRNA of gene} \xrightarrow[\text{dNTP}]{[\text{RT}]} \text{cDNA single stranded} \xrightarrow[\text{dNTP}]{[\text{DNA}-\text{P}]} \text{cDNA double stranded}$$

The isolated DNA is now treated with a particular endonuclease to give the 'sticky ends' to match those on the vector.

SYNTHESIS OF RECOMBINANT DNA

Copies of the gene now have to be made and this is done using a carrier molecule or *vector* which replicates in the host cell. The vectors most commonly employed are *plasmids*, which are extrachromosomal circular molecules of double-stranded DNA that occur naturally in bacteria and yeast. The plasmid is treated with the endonuclease and if the plasmid has a single restriction endonuclease target site then a

ISOLATE DNAs

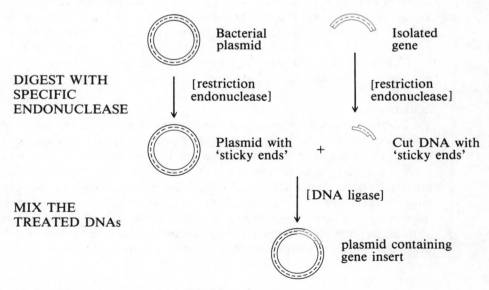

DIGEST WITH
SPECIFIC
ENDONUCLEASE

MIX THE
TREATED DNAs

Figure 11.10 The preparation of recombinant DNA.

linear molecule will be produced with 'sticky ends'. The fragments are now mixed with those produced by the treatment of the gene with the same endonuclease and under the right conditions the 'sticky ends' of the molecules from the two sources of DNA will loosely combine by hydrogen bonding to form a larger hybrid circular molecule. This combination is then made permanent by incubation with DNA ligase to give recombinant DNA (Fig. 11.10). The plasmid containing the required gene is now introduced into the host cell (e.g. *E. coli*) made transiently permeable to macromolecules.

GENE CLONING

Only a few bacteria will take up the recombinant plasmid and the problem then becomes how to select those bacteria which contain the foreign DNA from the vast majority of the bacteria which do not. This is done by using a plasmid which carries the genes that confer resistance to an antibiotic such as penicillin or tetracycline. If the treated *E. coli* are plated out on agar containing the antibiotic then only those bacteria with the antibiotic-resistant gene will grow. A single cell containing the plasmid will continue to divide until a *clone* of identical cells is formed on the agar. The 'resistant' clone of *E. coli* is then grown in culture to produce a large number of bacteria containing the recombinant DNA (Fig. 11.11). In this way large amounts of the gene may be produced and then isolated for further use or study. Alternatively, the culture may be grown under conditions which ensure that the foreign gene is expressed so that its protein product can be isolated.

FUTURE APPLICATIONS

Laboratory work in genetic engineering now includes bacteria other than *E. coli*, yeast and also plant and animal cells. The use of vectors such as bacteriophages, viruses and plasmids from other organisms is also being investigated. There are many technical problems to be overcome in this area of molecular biology but it seems likely that research in genetic engineering will lead to significant contributions in the fields of agriculture, biotechnology and medicine.

AGRICULTURE

The insertion of specific genes into plants could result in a series of 'tailor made' plants which are able to grow under extreme conditions of drought, temperature and high salt concentration in the soil. Another example is the possible addition to plants of genes from soil bacteria which are responsible for the production of natural insecticides or resistance to herbicides. Similarly, genes that code for nitrogen fixation might be incorporated into plants which would avoid the use of costly fertilizers. The potential is also there for improving the nutritional quality of plants, for example by introducing more of the amino acids methionine and lysine into plant proteins, many of which are deficient in one or other of these essential amino acids.

BIOTECHNOLOGY

Genetic engineering also offers the opportunity for producing fine chemicals such as

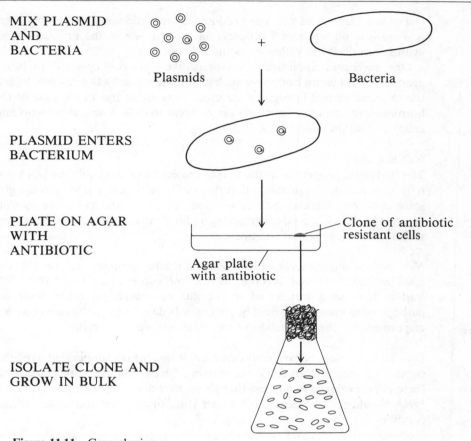

MIX PLASMID
AND
BACTERIA

Plasmids   +   Bacteria

PLASMID ENTERS
BACTERIUM

PLATE ON AGAR
WITH
ANTIBIOTIC

Clone of antibiotic
resistant cells

Agar plate
with antibiotic

ISOLATE CLONE AND
GROW IN BULK

Figure 11.11 Gene cloning.

drugs, vaccines and hormones on a large scale and at a reasonable cost. Large
quantities can be made and the end product can be realized without having to go
through a series of extractions and modifications of a primary product. The other
main advantage is the energy saved, since biological processes generally occur at
moderate temperatures and under relatively mild conditions which avoids the use of
powerful oxidants, reducing agents, acids and bases. The use of engineered micro-
organisms also raises the possibility of treating effluents and recycling waste to make
it less toxic and even to produce some useful compounds as end products.

MEDICINE
The insertion of genes to replace those which are missing or defective is at the moment
only a dream but one that hopefully may be realized in the not too distant future. If
this can be done then inherited diseases such as phenylketonuria and the glycogenoses
can be treated and cured completely.

The manufacture of vaccines produced by recombinant DNA technology however is a realistic objective and progress has been made in the production of vaccines against certain forms of hepatitis and herpes.

One successful application, which is likely to be expanded in future, is the production of human hormones such as insulin and growth hormone by bacteria. At the moment animal hormones are used in medicine but in the case of the peptide hormones, the primary structures are different to those found in humans and this can cause undesirable side effects.

POSSIBLE RISKS

The biological properties of the hybrid molecules of recombinant DNA are still not fully known and the possibility that they could be transferred to man has given rise to some concern. Working parties were set up by several countries to examine the possible hazards of genetic engineering and they all concluded that there are two areas which could be potentially dangerous:

New pathogenic bacteria *Escherichia coli* are normally present in the human intestine and accidental infection with *E. coli* containing a new hybrid DNA could lead to this being transferred to the gut bacteria or to other bacteria that are pathogenic to man. This would be particularly dangerous in the case of bacterial genes responsible for the production of toxins or antibiotic resistance.

Cancer The other hazard considered a risk was the possibility that hybrid molecules containing animal DNA could become transferred to humans and cause cancer. So far there is no evidence to suggest that these fears are real but work with recombinant DNA should only be carried out under close supervision and under stringent safety conditions.

12. Metabolic control

12.1 Regulation of enzyme activity: allosteric enzymes

General principles

METABOLIC REGULATION

Metabolism in living organisms is far from simple. There are a large number of metabolic pathways and each one has its own array of enzymes, substrates and cofactors so that many thousands of reactions are taking place at the same time. Metabolism is further complicated by the fact that many intermediates are common to more than one pathway so that several metabolic conversions are possible. Glucose, for example, can be broken down to give energy in glycolysis, stored as glycogen in glycogenesis or metabolized to give intermediates and NADPH for biosynthesis in the pentose phosphate pathway. Exactly how much glucose is metabolized by each route depends on the needs of the organism and its physiological state so that there is order among all this complexity. Another example of this order is the fact that different metabolic pathways are balanced so that there is no excess or deficiency of components needed for the synthesis of a particular molecule or structure. A good example of this is the synthesis of RNA and DNA where the supply of the purines and pyrimidines needs to be the same. All this shows that metabolism is integrated and also carefully regulated. Metabolism is largely controlled by increasing or decreasing the activity of key enzymes and this can be achieved in a number of ways. The most direct of these is the modulation of the catalytic activity by changes in the concentration of substrates, cofactors, activators and inhibitors. This is the most rapid form of metabolic control and is the one discussed in this section.

CONTROL POINTS OF METABOLIC PATHWAYS

Consider the following metabolic pathway in which A is metabolized to D in a series of reactions catalysed by the enzymes E_1, E_2 and E_3:

$$A \xrightarrow{[E_1]} B \xrightarrow{[E_2]} C \xrightarrow{[E_3]} D$$

The rate at which A is converted to D depends on the activities of the enzymes in the pathway. Control is exercised through a key enzyme which is usually the first in a pathway. If the activity of $E_2 > E_1$ then B is removed rapidly so that E_1 never attains equilibrium and the rate of formation of D depends on the activity of E_1. In other words the conversion of A to B, catalysed by the non-equilibrium enzyme E_1, is the *flux generating step*. Such control points on metabolic pathways are usually found to be the first reaction of a sequence or at a branch point.

Changes in the flux through the system depend on the degree of saturation of the enzyme with substrate. Those enzymes that are approaching saturation have zero-order kinetics and are virtually unaffected by changes in the concentration of substrate. However, enzymes that are far from saturation have first-order kinetics and their activity depends on the concentration of substrate (Section 6.2). This means that changes in the concentration of A do not affect the flux if E_1 is almost saturated but have a marked effect on the activity if E_1 is far from saturation.

FEEDBACK INHIBITION

Enzymes are sensitive to inhibitors as well as substrates and the rate of a metabolic pathway is often controlled by inhibitors. In some cases the enzyme is inhibited by one of the products of the reaction so that if a metabolite is not removed then its rate of formation is cut down. A good example of this is the inhibition of hexokinase (HK) by glucose-6-phosphate. This is extremely important in the cell since the inhibition of hexokinase by its product prevents the total removal of glucose by phosphorylation.

$$\text{glucose} + \text{ATP} \xrightarrow[\text{Mg}^{2+}]{\text{[HK]}} \text{G}-6-\text{P} + \text{ADP}$$

In many cases however enzymes are inhibited by the final product of a metabolic sequence and this compound is likely to have quite a different structure to the substrate or product of the enzyme. These enzymes therefore have specific binding sites for inhibitors which are quite distinct from the active site and are known as *allosteric enzymes* (Gk *allos* = other, *stereos* = solid).

Allosteric enzymes

The name comes from their ability to change shape when exposed to *effectors* such as inhibitors, activators or substrates. Such a change in shape, caused by one effector, has a marked effect on the binding of other effectors such as substrate so that their kinetics cannot be described by the classical Michaelis–Menten theory.

KINETICS

Michaelis–Menten enzymes give a plot of activity (v) against substrate (s) which is a rectangular hyperbola (Fig. 6.4) and can be described by the equation:

$$v = Vs/s + K_\text{m}$$

Allosteric enzymes on the other hand give a sigmoidal type of plot (Fig. 12.1) when v is plotted against s. This is because they have more than one binding site and the binding of each molecule of substrate effects the binding of subsequent molecules of substrate. This enhancement of binding is known as the cooperative effect. Activators

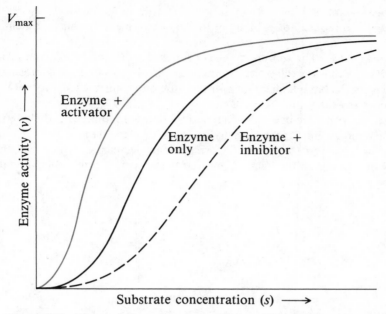

Figure 12.1   The variation of enzyme activity with substrate concentration for a allosteric enzyme.

are positive effectors and make the v against s plot less sigmoidal whereas inhibitors have the opposite effect (Fig. 12.1).

The kinetics are not straightforward and the binomial theorem and probability factors have to be used to derive the strict equation. However, a simplified derivation, which gives an approximate equation, has been obtained by Atkinson who made the following assumptions:

1. There are n binding sites for the substrate.
2. Equilibrium is rapidly attained between the enzyme and substrate.
3. The intermediates have only a transient existence.

Under these conditions the equation derived is:

$$v = V[s]^n/K + [s]^n$$

n   In theory n is the number of binding sites but in practice this is always less than this value since the above assumptions are not completely correct. The value of n therefore is a measure of the degree of cooperativity: the bigger the value of n the more sigmoid the curve. If n is 1 the curve is no longer sigmoid but a rectangular hyperbola.

K   This is not the same as K_m but is a complete steady-state constant. In the case of hyperbolic kinetics, when $K_m = s$, $v = V/2$ but for sigmoid kinetics when $K_m = s$, $v \neq V/2$.

ALLOSTERIC MODELS

There are two models for explaining the sigmoid kinetics of allosteric enzymes.

Concerted model The *concerted model* proposed by Monod, Changeux and Wyman basically says that the allosteric enzyme can exist in two distinct conformations. The relaxed form (R) has a high affinity for the substrate while the tense form (T) has a low affinity for the substrate.

These two forms are in equilibrium and inhibitors bind preferentially to the T form while activators bind preferentially to the R form. This causes a shift in the equilibrium between the R and T forms which decreases or increases the enzyme activity depending on the ligand bound. If the enzyme is a tetramer then this can be depicted as:

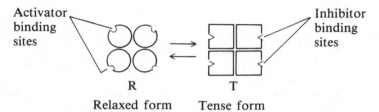

Activator binding sites

Inhibitor binding sites

R — T

Relaxed form   Tense form

Sequential model The *sequential model* proposed by Koshland is similar, in that two conformations are possible, but this time the proposal is that the subunits can exist in these two forms. The binding of substrate to one subunit causes a change in the conformation of that subunit. This in turn interacts with an adjacent subunit so that the next substrate molecule is bound more readily.

If the relaxed and tense subunits are shown as circles and squares then the sequential changes in the conformation of an allosteric enzyme which is a tetramer would be as follows:

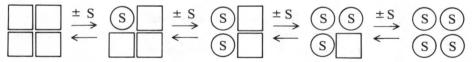

In the case of this model, activators are compounds which induce a change in the conformation of a subunit that is similar to that of the substrate while inhibitors have · the opposite effect.

Allosteric control of metabolic pathways

Metabolic control by the regulation of allosteric enzymes is very rapid and takes place in less than a second. Two examples of this type of control are given below.

THE SYNTHESIS OF PYRIMIDINES

A good example of allosteric control is the enzyme *aspartate transcarbamoylase* (ATC) of *Escherichia coli*. This catalyses the formation of *N*-carbamoylaspartate from

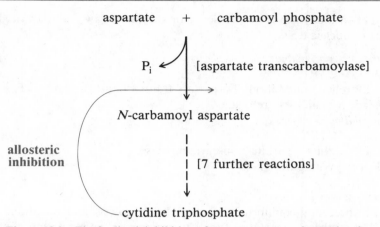

Figure 12.2   The feedback inhibition of aspartate transcarbamoylase by cytidine triphosphate.

aspartate and carbamoyl phosphate (Fig. 12.2) and is the first step in the synthesis of pyrimidines. Cytidine triphosphate (CTP), which is the final product of the pathway, controls the activity of ATC by feedback inhibition. There are six catalytic subunits and six regulatory subunits which bind the CTP. The purine ATP competes with CTP for the binding sites on the regulatory subunits and relieves the inhibitory action of CTP (Fig. 12.3). ATP therefore balances the supply of purine and pyrimidine

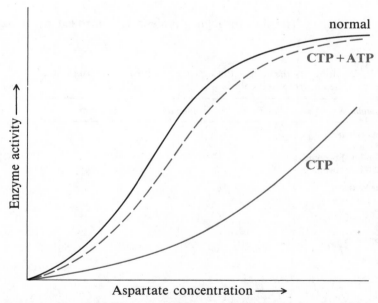

Figure 12.3   The inhibition of aspartate transcarbamoylase by CTP and its relief by ATP.

nucleotides so that there is no deficiency or surplus of raw materials needed for the synthesis of nucleic acids.

1. *If purines > pyrimidines*
   (a) ATP > CTP
   (b) ATP relieves inhibition of aspartate transcarbamoylase by CTP
   (c) More pyrimidines are made
2. *If pyrimidines > purines*
   (a) ATP < CTP
   (b) CTP inhibits aspartate transcarbamoylase
   (c) Fewer pyrimidines made

THE CONTROL OF GLYCOLYSIS

There are three non-equilibrium reactions in glycolysis and metabolic control is exercised at all these steps. One of the key enzymes in the regulation of glycolysis is *phosphofructokinase* (PFK) which catalyses the irreversible phosphorylation of fructose-6-phosphate by ATP to give fructose-1,6-bisphosphate:

$$\text{fructose-6-phosphate} + \text{ATP} \xrightarrow[\text{Mg}^{2+}]{\text{[PFK]}} \text{fructose-1,6-bisphosphate} + \text{ADP}$$

This reaction is the first committed step of glycolysis proper since the earlier hexose phosphates can all be metabolized by other pathways. It is an ideal example of allosteric control since the enzyme is affected by a number of compounds, some of which activate while others inhibit (Table 12.1).

Adenine nucleotides   The flux through glycolysis is controlled by changes in the

Table 12.1 Compounds that affect the activity of phosphofructokinase a key enzyme in the control of glycolysis

Enzyme modulator	Substrate	Inhibitor	Activator
Fructose phosphates			
Fructose-6-phosphate	+		+
Fructose-1,6-bisphosphate			+
Fructose-2,6-bisphosphate			+
Adenine nucleotides			
ATP	+	+	
ADP			+
AMP			+
Other compounds			
Citrate		+	
Ions			
Mg^{2+}			+
NH_4^+			+
P_i			+

concentration of the adenine nucleotides which reflect the energy requirements of the organism.

Adenosine triphosphate (ATP) is one of the substrates of PFK but it is also an allosteric inhibitor. This seems contradictory but the catalytic site has a higher affinity (lower K_m) than the regulatory site so that the enzyme activity at first rises with increasing concentration of ATP then falls when there is sufficient ATP for the needs of the organism.

Adenosine monophosphate (AMP) relieves the inhibition by ATP so that when the level of AMP rises there is a need for more ATP and the flux through glycolysis is increased. This is a particularly sensitive signal because the level of AMP varies as the square of the concentration of adenosine diphosphate (ADP). This is due to the very active enzyme *adenylate kinase* (AK) which catalyses the formation of ATP and AMP from two molecules of ADP:

$$ADP + ADP \xrightleftharpoons{[AK]} ATP + AMP$$

$$K_{equ} = [ATP][AMP/[ADP]^2$$

$$[AMP] = K_{equ}[ADP]^2/[ATP]$$

Citrate The glycolytic flux is also controlled in muscle but not in liver by the level of citrate which enhances the inhibitory action of ATP on PFK. The accumulation of citrate may be due to a reduction in the use of Krebs'-cycle intermediates for biosynthesis, a rise in the oxidation of fatty acids or an increased rate of glycolysis. In all these cases there is a need to avoid a build-up of citrate and this is achieved by inhibition of PFK.

This is particularly useful during starvation when acetyl-CoA from the oxidation of fatty acids combines with oxaloacetate to give citrate which then spares the utilization of glucose by inhibiting PFK.

The decrease of citrate and ATP due to the synthesis of fatty acids has the opposite effect so that PFK works faster to supply the acetyl-CoA needed.

Pasteur effect Allosteric regulation of PFK is the most likely explanation of the *Pasteur effect*. This was a phenomenon discovered by Louis Pasteur who observed that the rate of fermentation of glucose was sharply increased in the absence of oxygen. Another way of looking at this is to say that fermentation is inhibited in the presence of oxygen and this can be explained by the increased levels of citrate in aerobic metabolism.

12.2   Hormones: the body's chemical messengers

The control of metabolism by the regulation of allosteric enzymes is brought about by changes in the concentration of metabolites inside the cell. In the case of multicellular organisms, metabolic control is also exerted by changes in the concentration of *hormones* produced outside the cell. Hormones are chemical messengers which are synthesized by the *endocrine glands* and transported in the circulation to their target cells. The effect of hormones on the physiology of the animal is well documented but in many cases their mechanism of action at the molecular level is still not fully understood. However, hormones can be divided into two general categories in terms of their primary effect.

The first group, which include the steroid and thyroid hormones, enter the cell and exert their activity in the nucleus. They control transcription which affects enzyme activity by increasing or decreasing the rate of synthesis and degradation of the enzyme proteins. This is a long-term form of metabolic control which takes place over several hours or days.

The second group, which include the catecholamines and the polypeptide hormones, do not enter the cell but elicit their response at the cell membrane. They do this by changing the permeability of the membrane to specific metabolites or by inducing the formation of another signal which increases or decreases the activity of cellular enzymes by covalent modification of the enzyme protein. This takes place in seconds or minutes, which is quite fast but not as rapid as allosteric control where responses are measured in fractions of a second. Examples of hormone action in this section are taken from this second category since more is known about their action at the molecular level than hormones which control gene transcription.

The effect on membranes

HORMONE RECEPTORS

The catecholamines and peptide hormones bind to specific receptors located on the cell membrane of their target cells. These receptors are proteins or glycoproteins which have a high *specificity* and a high *affinity* for the hormone. The binding constants are quite high but this is necessary because of the extremely low concentration of hormones in the circulation (10^{-9}–10^{-12} M). The biological response does not appear to be related to their binding affinity but in many instances does depend on the number of receptors occupied. Some hormones show a threshold effect, in that a certain number of receptors needs to be occupied before any response is obtained. In the case of other hormones, a maximal response is realized when only a fraction of the receptors are occupied. Changes in the number of receptors have been shown to be an important way of regulating the activity of some hormones.

MEMBRANE TRANSPORT

Insulin is a good example of a hormone which affects membrane transport. It causes a whole host of changes in the metabolism of the animal which can be summed up by saying that insulin stimulates anabolic processes and suppresses catabolic processes.

The most obvious of these is its action in conserving fat and promoting the use of glucose. Many of the details of how it does this are still not fully understood but its action on the transport of glucose and amino acids across membranes is well documented. The increased secretion of insulin which follows a carbohydrate meal promotes the uptake of glucose from the blood into muscle and adipose tissue. It does this by binding to its receptor on the cell surface and this brings about a change in the cell membrane which stimulates the transport of glucose, amino acids and ions into the cell.

The binding of a hormone to its receptor is similar in many ways to the interaction of a substrate and an enzyme and the process can be defined in terms of a dissociation constant (K_m) and a maximum effect (V_{max}). In the case of glucose transport, insulin increases the V_{max} but has no effect on the K_m. However, insulin does not affect the transport of glucose into the brain because this organ must be able to use glucose even at a low blood concentration when the secretion of insulin is reduced.

SECOND MESSENGERS

Many hormones act by stimulating the production of a *second messenger* and it is this which triggers the metabolic response in the cell. The second messenger is therefore a mediator which transmits the membrane signal produced by the hormone to the interior of the cell. A number of second messengers have been discovered including Ca^{2+}, prostaglandins, diacylglycerol and cyclic guanosine monophosphate (cGMP) but the best known example is 3′,5′-cyclic adenosine monophosphate. This is usually abbreviated to cyclic-AMP or cAMP and is the mediator for several hormones including adrenaline, glucagon and thyroid stimulating hormone (TSH). The hormone binds to its receptor on the outside of the cell membrane and stimulates adenylate cyclase (AC) on the inside of the membrane which converts ATP to cAMP and pyrophosphate (PP_i) (Fig. 12.4). The reaction is close to equilibrium but effectively operates in one direction only due to the action of an active pyrophosphatase which rapidly removes one of the products of the reaction.

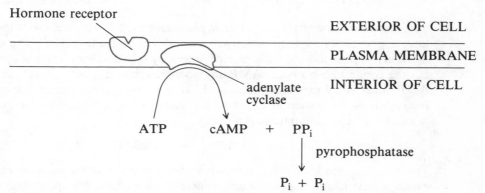

Figure 12.4 The production of the second messenger cyclic-AMP.

Cyclic-AMP is metabolized to AMP by a phosphodiesterase so the level of cAMP in the cell is governed by the balance of adenylate cyclase and phosphodiesterase. Caffeine in coffee and tea and theophylline found in some chocolate bars increase the level of cAMP by inhibiting phosphodiesterase. This leads to the mobilization of metabolic fuels and partly explains why these compounds are effective stimulants.

The covalent modification of enzymes

The increased levels of cAMP in the cell lead to a series of changes involving the covalent modification of certain enzymes by phosphorylation. The phosphorylated and non-phosphorylated forms of the enzymes catalyse the same chemical reaction but at quite different rates and in most cases one of the forms is virtually inactive.

In the case of enzymes the addition or removal of a phosphate group means that the activity can be switched on or off depending on the presence of hormones. This is a rapid form of metabolic control and this principle is illustrated with two examples of enzymes involved in the metabolism of glycogen.

PHOSPHORYLASE

A key enzyme in the breakdown of glycogen is phosphorylase and this can be activated by phosphorylation. Serine residues on the enzyme are phosphorylated by ATP in a reaction catalysed by *phosphorylase kinase* (PhK) and this converts the inactive b form of phosphorylase to the active a form:

serine residue

PHOSPHORYLASE b

phosphorylated
serine residue

PHOSPHORYLASE a

This reaction can be reversed by *protein phosphatase* (PrP) which converts the active a form of phosphorylase to the inactive b form of the enzyme (Fig. 12.5). The formation of the active phosphorylase a is brought about by raised cellular levels of cyclic-AMP but the connection between cAMP and phosphorylase a is not a direct one. Cyclic-AMP binds to the regulatory subunit (R) of the allosteric enzyme cAMP-dependent *protein kinase* or *protein kinase A* which leads to the activation of the enzyme by relasing the active catalytic subunits (C):

$$R_2C_2 + 4cAMP \longrightarrow R_2(4cAMP) + 2C$$

The active protein kinase A then catalyses the phosphorylation of serine residues in *phosphorylase kinase* to give the active form of the enzyme. Phosphorylase kinase in turn catalyses the phosphorylation of phosphorylase b to give the active a form. This

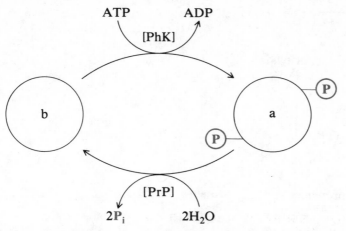

Figure 12.5   Two forms of muscle phosphorylase. The predominant form in resting muscle is b and during exercise this is converted to the active a form. (PhK = phosphorylase kinase, PrP = protein phosphatase.)

sequence of events appears complicated but such a cascade means that the original membrane signal can be greatly amplified (Fig. 12.6) so that relatively small amounts of a hormone can produce quite a large effect.

ADRENALINE

A good example of this is the hormone adrenaline which is released from the adrenal glands when the animal is confronted with an emergency situation so that it either fights or runs. In either case there is a need for the rapid breakdown of glycogen in the muscles and only a few molecules of adrenaline are needed for the rapid mobilization of large numbers of molecules of glucose-1-phosphate from glycogen ready for oxidation. However, there would be little point in mobilizing glycogen until it is actually needed when the muscles go into action. Stimulation of the muscles by the nerve impulse causes the release of Ca^{2+} and this not only causes muscular contraction but also activates phosphorylase kinase. The nervous impulse therefore causes the full potential for the mobilization of glycogen by adrenaline to be actually realized. Ca^{2+} and cAMP therefore act synergistically to stimulate the system completely. Adrenaline also mobilizes liver glycogen but in this case the glucose-1-phosphate produced is metabolized to glucose-6-phosphate, then glucose which is released into the bloodstream.

Glucagon   Adenylate cyclase is also associated with receptors to the pancreatic hormone glucagon in the liver but not in muscle. Glucagon therefore mobilizes liver glycogen and raises the blood glucose but has no effect on the glycogen of muscle. Hormone receptors to both adrenaline and glucagon are also found in adipose tissue but in this case the raised cyclic-AMP causes phosphorylation of *triacylglycerol lipase,*

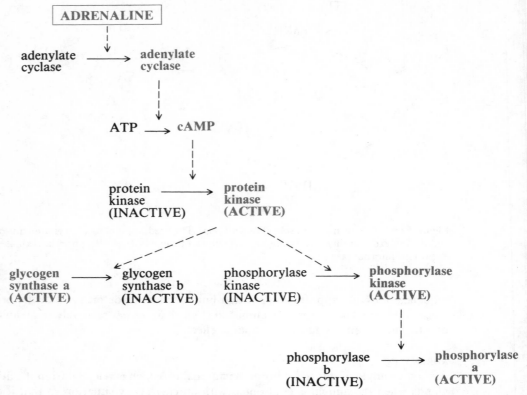

Figure 12.6 The control of the activities of phosphorylase and glycogen synthase by adrenaline.

which leads to the mobilization of fat and explains how these hormones stimulate lipolysis.

GLYCOGEN SYNTHASE

The synthesis and breakdown of glycogen is determined in large part by the balance between the activity of the synthetic enzyme glycogen synthase and the degradative enzyme phosphorylase (Fig. 12.6). Therefore, in order to achieve maximum efficiency, glycogen synthase needs to be 'switched off' at the same time that glycogen breakdown is 'switched on' and this is done by phosphorylation. Glycogen synthase is similar to phosphorylase in that it exists in an active (a) and an inactive (b) form but in this case it is the non-phosphorylated form which is active and the phosphorylated form which is inactive. The sequence of events that leads to adrenaline or glucagon stimulating phosphorylase therefore has the opposite effect on glycogen synthase (Fig. 12.6). Phosphorylation and dephosphorylation therefore control the balance between the breakdown and the synthesis of glycogen.

The above example illustrates the general principle that for degradative reactions the active enzyme is the phosphorylated form while the inactive enzyme is the non-phosphorylated form. The opposite is true for synthetic reactions where the active enzyme is the non-phosphorylated state and the inactive enzyme is the phosphorylated state.

Covalent modification of proteins by phosphorylation is very widespread and appears to be involved in many other important biological activities as well as in the control of enzyme activity. These include fertilization, memory, membrane transport, cardiac function, control of the cell cycle and the action of neurotransmitters.

12.3   Metabolic integration: oxidizable fuels in starvation

The integration of metabolism is a highly complex subject and there are many areas that are still not fully understood. However, quite a lot is known about the pathways involving the metabolic fuels and this aspect of metabolism is used to illustrate some of the principles of metabolic integration.

Animals can oxidize a number of metabolic fuels including glucose, fatty acids, glycerol, ketone bodies and amino acids, and the relative contribution of each of these fuels to energy production depends on the physiological state of the animal. The changes in the utilization of these metabolic fuels and their stores of glycogen, triglyceride and protein during starvation are discussed in this section and the disruptive effect of *diabetes mellitus* on metabolic integration in Section 12.4 which follows. Starvation occurs when there is an inadequate supply of food in the form of calories and protein available to the animal. Under these circumstances the animal has to draw on its own reserves of food in order to survive and marked changes occur in the balance of the fuels that are oxidized by the various tissues of the body. The relative contribution of the two main fuels of carbohydrate and lipid is reflected in the blood chemistry and Fig. 12.7 shows the changes that occur in the blood glucose, fatty acids and ketone bodies during prolonged starvation in man.

Fasting (up to 24 hours)

With no food being digested the flow of nutrients from the gut ceases and the animal begins to draw on its own reserves of food.

GLYCOGEN AND GLUCOSE

During the first day without food, glucose continues to be the main fuel used by the tissues and this causes a sharp fall in blood glucose from about 6 to 4 mM. There is only enough glucose in the blood for a few hours' supply for the tissues so that liver glycogen is broken down to glucose in order to maintain the blood concentration at an acceptable level.

The trigger for the mobilization of liver glycogen is the lower blood sugar which signals the pancreas to increase the secretion of glucagon and reduce that of insulin. The raised levels of the hormone glucagon then activate phosphorylase and mobilize glycogen as discussed in Section 12.2. However, this reserve of glucose is limited and liver glycogen becomes depleted after about 24 hours of normal activity.

The blood glucose level is also maintained by gluconeogenesis (Section 9.1) which is stimulated by the increased secretion of glucagon and the glucocorticoids and rises to a maximum rate at the end of this period. Lactate from muscle glycogen and erythrocytes, glycerol from the hydrolysis of triacylglycerols and amino acids are all converted to glucose by the liver.

TRIGLYCERIDES AND FATTY ACIDS

Insulin inhibits and glucagon stimulates lipolysis so the increased ratio of glucagon/insulin begins to mobilize the triglycerides from adipose tissue. There is also

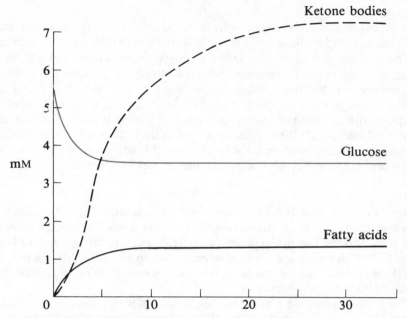

Figure 12.7   Changes in the blood glucose, fatty acids and ketone bodies during starvation.

a rise in the secretion of growth hormone and this also stimulates lipolysis. The resulting fatty acids then provide an alternative fuel to that of glucose and tissues such as heart and skeletal muscle begin to oxidize fatty acids in preference to glucose. This leads to a rise in acetyl-CoA and citrate which inhibits phosphofructokinase and reduces the flux through glycolysis. This has the effect of sparing glucose for those tissues such as brain and erythrocytes which are dependent on this metabolic fuel.

KETONE BODIES
Peripheral tissues also begin to oxidize ketone bodies but at this stage their use as an alternative fuel is only of minor importance compared to glucose and fatty acids and this is reflected in the modest rise observed in their concentration in the circulation.

PROTEINS AND AMINO ACIDS
Some proteolysis of muscle proteins also takes place and most of the resulting amino acids which are released into the blood are taken up by the liver and converted to glucose by gluconeogenesis.

Starvation (up to 1 month)
During the next period the trends started early on develop and stabilize. The synthesis of lipogenic enzymes is reduced while enzyme induction increases the activity of those involved in gluconeogenesis and the urea cycle.

GLUCOSE

The glycogen stores have now been depleted and the blood glucose concentration continues to fall for about another week. At the end of this period it stabilizes around 3.5 mM in spite of a continuing fall in gluconeogenesis which does not stabilize until the end of the month. The reason why the blood glucose does not fall in line with gluconeogenesis but maintains a steady blood level is due to the reduction in the utilization of glucose by the tissues. The brain in particular must continue to be supplied with glucose but during prolonged starvation it becomes less dependent on this metabolic fuel. At the beginning of the fast the brain needs about 100 g of glucose per day but after starvation for several weeks this falls to about 35 g per day, as the brain begins to use ketone bodies as a fuel as well as glucose.

FATTY ACIDS

There is a continued rise in lipolysis and the oxidation of fatty acids and this is reflected in the rise in the circulating level of fatty acids. This reaches a plateau after one week and the rise in the plasma fatty acids mirrors the fall in the blood glucose (Fig. 12.7). Fatty acids are oxidized in preference to glucose by tissues that are able to use them (e.g. muscle) but the brain is unable to oxidize fatty acids since they cannot cross the blood–brain barrier.

The breakdown of fatty acids by β-oxidation in liver gives rise to acetyl-CoA and there is more acetyl-CoA produced than can be oxidized in the Krebs' cycle. This is because there is a relative deficiency of oxaloacetate needed for the oxidation of acetyl-CoA due to its conversion to glucose in gluconeogenesis. The liver then converts these (spare) acetyl residues to ketone bodies (Section 8.4) and these are used by the brain and other tissues but not by liver.

KETONE BODIES

During this second stage of starvation from one day to one month there is a dramatic increase in the ketone bodies in the blood (Fig. 12.7). Their concentration increases from only trace amounts of less than 0.03 mM to a plateau of about 7 mM after several weeks without food. Ketone bodies are soluble compounds unlike fatty acids which can only be transported in the blood plasma bound to albumin. Ketone bodies are therefore a convenient way of delivering a high concentration of a lipid fuel to the tissues. This high concentration also means that reasonable quantities can enter the cells of the brain where they are oxidized to provide energy. There is no change in the concentration of enzymes in the brain capable of oxidizing ketone bodies and their increased use by the brain is entirely due to the high circulating levels in the blood. Ketone bodies are also used by extrahepatic tissues and during starvation up to a quarter of the energy needs of heart and skeletal muscle come from the oxidation of acetoacetate and 3-hydroxybutyrate.

The disadvantage of ketone bodies is that two are acids and high concentrations cause metabolic acidosis. This leads to an increased activity of glutaminase and glutamate dehydrogenase in the kidney and an increased excretion of NH_4^+ in the urine (Section 10.1). These two reactions acting sequentially actually generate $2NH_4^+$

which is excreted in the urine. If the further metabolism of 2-oxoglutarate is considered we can see how very effective this is at removing the excess H^+.

$$
\begin{array}{c}
COO^- \\
| \\
(CH_2)_2 \\
| \\
C{=}O \\
| \\
COO^-
\end{array}
\quad + \quad 2H^+ \quad + \quad 5H_2O \quad \xrightarrow[\text{oxidation}]{\text{total}} \quad 5CO_2 + 8H_2O
$$

$$
2
\begin{array}{c}
COO^- \\
| \\
(CH_2)_2 \\
| \\
C{=}O \\
| \\
COO^-
\end{array}
\quad + \quad 4H^+ \quad + \quad 2O_2 \quad \xrightarrow{\text{gluconeogenesis}} \quad C_6H_{12}O_6 + 4CO_2
$$

MUSCLE PROTEIN

The increased glucocorticoids stimulate protein catabolism and muscle proteins are broken down to amino acids which are partially metabolized by the muscle. However, some are released into the circulation and are taken up by the liver and kidney where they are converted to glucose. Two of the amino acids released from muscle more than others are alanine and glutamine. Alanine is a very good glucogenic amino acid and some of the glutamine is oxidized by the cells of the gut and the rest is converted to alanine.

FAMINE (GREATER THAN 1 MONTH)

Prolonged starvation for more than one month occurs with famine which sadly for centuries has been the lot of much of humanity.

Steady state At the end of a month without food a steady state is reached in the circulating levels of glucose, fatty acids and ketone bodies (Fig. 12.7). Physiological changes also occur which maximize the chance of survival and eke out the remaining reserves of food. An increased excretion of thyroid hormones, particularly a modified form of triiodothyronine (T_3), lowers the basal metabolic rate so that less energy is needed. This and other factors mean that those suffering from prolonged starvation become less active. Psychological changes also take place and starving individuals become apathetic and withdrawn.

Death Clearly this state of affairs cannot continue indefinitely and ends in death unless food becomes available. Death must occur when the triglyceride stores, together with the available protein, are all gone but it usually takes place before this. The circulating level of antibodies and the number of white blood cells become so low that the defences of the body become very weak. This means that invading organisms can get a hold and a common cause of death in starvation is pneumonia. Death by starvation is preventable since there is sufficient food in the world to feed everyone. In spite of modern science and global communications more people die of famine than are killed in war. The answer to this terrible problem is clearly one of political will rather than agricultural science.

Summary

The changes in the circulating level of glucose, fatty acids and ketone bodies during starvation are shown in Fig. 12.8. The direction of the changes in these fuels and hormones is also shown in Table 12.2 which compares the fed and the fasting state. Finally, the movement of the metabolic fuels between the tissues is given in Fig. 12.8.

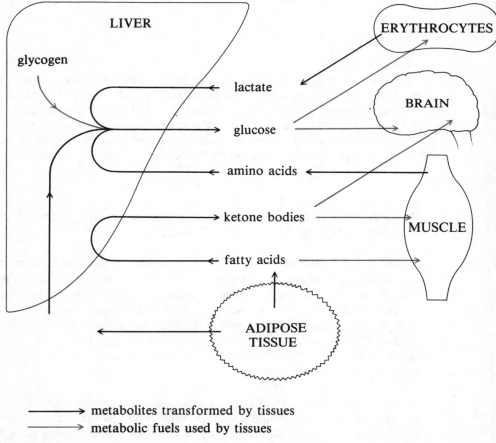

$\longrightarrow$ metabolites transformed by tissues
$\longrightarrow$ metabolic fuels used by tissues

Figure 12.8   Metabolic integration of body tissues in starvation.

Table 12.2 Metabolic changes in the fed and fasted states

Metabolite or hormone	Fed	Fasting
Blood glucose	↑ (8 mM)	↓ (4 mM)
Plasma insulin	↑ (80 μU/ml)	↓ (10 μU/ml)
Glycogen synthesis	↑ active	— stops
Lipogenesis	↑ active	— stops
Plasma fatty acids	↓ (0.03 mM)	↑ (1.4 mM)
Plasma ketone bodies	↓ (0.02 mM)	↑ (7 mM)
Plasma glucagon	↓ (70 pg/ml)	↑ (150 pg/ml)
Glycogen breakdown	— none	↑ active
Lipolysis	— none	↑ active
Proteolysis	— low	↑ active
Heart and muscle fuel	glucose	fatty acids amino acids ketone bodies
Brain fuel	glucose	glucose ketone bodies

12.4   Metabolic integration in disease: diabetes mellitus

Diabetes mellitus

The words diabetes and mellitus come from two Latin words meaning 'a siphon' and 'sweet'. The name therefore describes one of the main symptoms of the disease, namely the passing of a large volume of urine which has a characteristic sweet taste. This condition is frequently called diabetes for convenience but it should not be confused with *diabetes insipidus*, which is a rare and totally unrelated disease, apart from the common symptom of the excretion of a large volume of urine.

OCCURRENCE AND SYMPTOMS
Diabetes mellitus is a common disease affecting large numbers of people throughout the world.

Incidence   It is difficult to arrive at an accurate figure for the incidence of diabetes since random population surveys have always revealed large numbers of previously undiagnosed cases. In the United Kingdom there are probably about 1 million diabetics and in the United States about ten times this number. The disease is treatable but in spite of this it is still a major cause of blindness and death in many countries.

Glucosuria   The most obvious symptom of overt diabetes is the presence of the sugar glucose in the urine; a condition known as *glucosuria*. This gives the urine a sweet taste and is a feature that was reported many years ago by the ancient Egyptians. Today, glucosuria is no longer detected by taste but by chemical means and a useful and rapid way of screening large numbers of urine samples is by means of a 'dip stick' incorporating the enzyme glucose oxidase (GO). This enzyme catalyses the oxidation of glucose to gluconic acid with the formation of hydrogen peroxide. Peroxidase (P), which is also incorporated into the test stick, catalyses the breakdown of the hydrogen peroxide and the oxygen released then reacts with a dye to give a coloured product:

$$\beta\text{-}D\text{-glucopyranose} + H_2O + O_2 \xrightarrow{\text{[GO]}} D\text{-gluconic acid} + H_2O_2$$

$$H_2O_2 \xrightarrow{\text{[P]}} H_2O + \tfrac{1}{2}O_2$$

$$\tfrac{1}{2}O_2 + \text{dye} \longrightarrow \text{coloured product}$$

The stick, which is easy to use, is simply dipped into the urine sample and changes colour if glucose is present. This is a useful diagnostic aid since glucose is absent from the urine of healthy individuals.

Hyperglycaemia   Glucose appears in the urine because of a high concentration of the sugar in the blood. This is called *hyperglycaemia* and in very serious cases the blood glucose can reach values ten times that found in the blood of healthy individuals.

Change in the blood glucose following the oral intake of glucose is a useful diagnostic test for diabetes and this is discussed later.

Ketonuria   If the diabetes is severe there is a high concentration of ketone bodies in the blood and also the urine. Ketone bodies are normally absent from the urine of healthy subjects and *ketonuria* is another diagnostic indicator of the disease.

THE PRIMARY LESION
These changes are brought about by a deficiency of insulin or a reduced sensitivity to the action of the hormone.

Type I diabetes   In this case there is a partial or complete destruction of the B-cells of the pancreas which means that little or no insulin is produced. This is the *insulin-dependent* or *Type I diabetes* and is the more serious though less common form of the disease. It is also known as *juvenile-onset diabetes* since it commonly appears in children and young adults although other age groups are affected. The reason why the B-cells are damaged and cease to function is not known but may be connected with a previous virus infection or exposure to an environmental toxin. There is also a genetic element which predisposes certain individuals to develop the disease although diabetes is not inherited directly.

Type II diabetes   In the other type of diabetes, insulin is present but the target cells are less sensitive to the action of the hormone. This is due to a fall in the number of receptors, a change in their structure or more rarely their blockage by antibodies. This form of the disease is called *insulin-independent* or *Type II diabetes* and is generally less serious but more common that the Type I form. It is also known as *maturity-onset diabetes* since it is mostly older people that are affected.

SECONDARY COMPLICATIONS
There are a number of secondary complications associated with the disease and diabetics are more likely to suffer from coronary heart disease or a stroke than non-diabetics. Diabetics are also very prone to cataract formation, damage to the retina (retinopathy), the degeneration of nerves (neuropathy) and kidney damage caused by the thickening of the basement membrane (nephropathy). Some of these complications may be due to hyperglycaemia which increases the glycosylation of proteins. For example, blood platelets become sticky and therefore more liable to form a clot and glycosylation of haemoglobin increases its oxygen affinity so that some tissues may become hypoxic. Other problems could be due to the insulin which is given to keep the disease in check. The hormone is known to stimulate the growth of smooth muscle and this affects the structure of the capillaries and therefore the flow of blood to the tissues.

Metabolic changes in uncontrolled diabetes
Diabetes causes many changes in metabolism and the most obvious one is the marked shift in the utilization of glucose and lipid fuels (Fig. 12.9). Insulin basically promotes

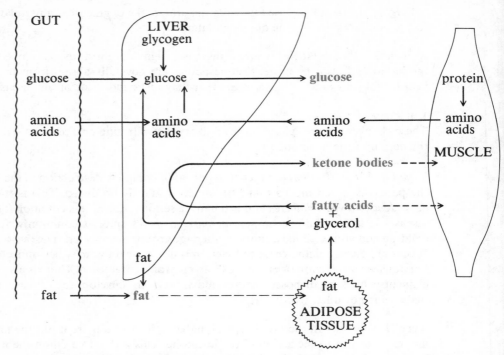

Figure 12.9 Metabolic interrelationships in Type I diabetes mellitus (metabolic fuels accumulating in the blood are shown in red).

the use of glucose and conserves fat so that during diabetes glucose is conserved and fat is mobilized. In diabetes therefore there is a fall in the oxidation of carbohydrates and an increase in that of fat.

GLUCOSE METABOLISM

One of the main features of diabetes as mentioned already is the high circulating level of glucose in the blood. This is because glucose is absorbed from the gut but is not taken up by the tissues or converted into fat so that it remains in the circulation. The situation is exacerbated by gluconeogenesis which is stimulated by the increase in the glucagon/insulin ratio brought about by the reduced secretion of insulin. The high blood glucose however is of little value since it cannot be used by the insulin-dependent tissues. In diabetes therefore there is starvation in the midst of plenty or to paraphrase the words of the 'Ancient Mariner':

'Glucose, glucose everywhere, Nor any a mole to use.'

A useful way to test for diabetes and other impairments in carbohydrate metabolism is to give an oral dose of glucose (1 g/kg) in water to a subject who has fasted overnight and to follow the change in the concentration of blood glucose over the next three hours. This is known as the *glucose tolerance test* and the results of such an experiment on a normal individual and a diabetic are shown in Fig. 12.10. The blood glucose concentrations quoted are those for capillary blood obtained from a finger prick. If venous blood is used the blood glucose is less by about 0.6 mM during fasting and 1.4 mM at the peak.

Normal glucose tolerance In a healthy subject the fasting blood glucose lies within the range 3.6–6.0 mM: following the ingestion of glucose this rises to a maximum of about 7.8 mM after 30–60 min. The pancreas responds to this rise in the blood sugar by increasing the secretion of insulin which promotes the uptake and utilization of glucose in a number of tissues. To begin with the rate of glucose absorption exceeds the rate of uptake but after the first hour the reverse is true and the blood glucose falls. The secretion of insulin lags behind the glucose so that the insulin concentration remains high even when the glucose is falling rapidly. The effect of this is to cause the blood glucose at 2 h to drop below the fasting level and that of the 3 h sample. This transient low blood sugar is the reason why there is a tendency to nod off a short time after a meal, a situation that is made worse if alcohol is consumed at the same time since this inhibits gluconeogenesis.

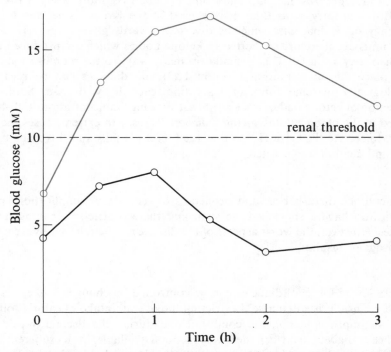

Figure 12.10 Results obtained from glucose tolerance tests on a normal individual (black) and a diabetic (red).

Glucose tolerance in diabetes Persons suffering from diabetes show a diminished glucose tolerance curve similar to that shown in Fig. 12.10. The actual values of the blood glucose depend on the severity of the disease but the general shape of the curve is the same. To begin with the fasting blood sugar is high and this rises quite sharply after the ingestion of the glucose reaching a peak later than that found with normal subjects. There is usually a fall in the blood glucose but this is small and the concentration is still abnormally high even after 3 h.

If the blood glucose concentration rises above the *renal threshold* which is about 10 mM, then the capacity for the kidney to reabsorb glucose is exceeded and glucose appears in the urine. Water accompanies the glucose and ions and this is the reason why diabetics pass large volumes of urine and become dehydrated.

LIPID METABOLISM

Insulin has a number of effects on lipid metabolism and these actions are reversed in diabetes when the normal delicate hormonal balance is disturbed. Insulin, for example, stimulates the conversion of glucose to fat but this is reduced in diabetes. Insulin also inhibits the release of fatty acids from the triglyceride stores in adipose tissue by blocking the action of cyclic-AMP. In diabetes this restraint is lifted and the cAMP produced by the action of other hormones stimulates the hormone-sensitive lipase and mobilizes the fat. Unfortunately there is more acetyl-CoA from the breakdown of fatty acids than can be used in the Krebs' cycle, due to a relative deficiency of oxaloacetate which is diverted towards gluconeogenesis. The excess acetyl units are therefore converted to ketone bodies which are produced at a faster rate than they are used. This is similar in many ways to the mobilization of fat that takes place during starvation (Section 12.3) and diabetes can be regarded as a pathologically extreme form of starvation caused by disease. Sometimes the production of ketone bodies is so great that very high concentrations (30–40 mM) are reached in the blood plasma giving a *diabetic ketosis*. In extreme cases the kidney is unable to secrete the extra H^+ so that the blood pH falls and this leads to acidosis, coma and death if uncorrected.

Treatment

The metabolic disturbances and proneness to secondary complications mean that diabetics do have a shortened life span and there is strictly speaking no cure for diabetes. However, the worst aspects of the disease can be relieved by treatment with diet, drugs or insulin.

DIET

The less severe forms of diabetes can be controlled by changes in the diet provided that some insulin is still active. Restricting the overall intake of calories often brings about an improvement in the condition and during the Second World War the incidence of diabetes fell when there was less food available. Many sufferers from Type II diabetes are obese and a reduction in their weight increases the number of insulin receptors present on the target cells. The same effect is obtained by exercise and a general increase in physical activity. Further improvements can be achieved by changing the balance of carbohydrate and fat in the diet and replacing fat containing saturated fatty acids with fat rich in polyunsaturated fatty acids. The introduction of slowly digested carbohydrate in the form of fibre from fruit and legumes is also beneficial. The fibre retards the rate of digestion by enzymes which smoothes out the rise in the blood glucose and the subsequent release of insulin. Sucrose on the other hand is absorbed very rapidly and gives rise to sharp fluctuations in the secretion of insulin. The intake of sucrose is therefore best reduced or cut out of the diet altogether.

DRUG THERAPY

Treatment with drugs is also useful provided some of the B-cells are still functional and the sulphonylurea drugs such as *tolbutamide* act by stimulating the B-cells to produce more insulin.

$$_3HC—\langle C_6H_4 \rangle—SO_2—NH—CO—NH—(CH_2)_3CH_3$$

1-butyl-*p*-tolylsulphonylurea (tolbutamide)

INJECTION WITH INSULIN

Only about 10 per cent of diabetics suffer from Type I diabetes and in about half of these cases the disease can be controlled by diet and drugs. The other half however need to inject insulin for the rest of their lives. The insulin does not cure the disease but does correct the metabolic disturbances by promoting the use of glucose and restricting fat metabolism to reasonable levels. Ever since its discovery by Banting and Best in 1921 insulin has saved many lives and improved the quality of life for millions of diabetics throughout the world. At the moment the insulin used is of animal origin and the slightly different structure compared to human insulin can produce undesirable side effects. It is hoped that human insulin produced by genetic engineering will soon be available on a large scale and this should overcome these problems.

Finally, this section, showing one of the many practical applications of biochemical knowledge, is probably a good note on which to end this book. I hope my readers have enjoyed this introduction to biochemistry and will go on to learn more about this fascinating subject.

Appendix

Biochemical abbreviations

The following abbreviations are those widely accepted in the biochemical literature and are often used without definition.

Abbreviation	*Name*
ATP	adenosine-5'-triphosphate
ADP	adenosine-5'-diphosphate
AMP	adenosine-5'-monophosphate
ATPase	adenosine triphosphatase
CoA (CoA-SH)	coenzyme A
acetyl CoA	acetyl coenzyme A
acyl CoA	acyl coenzyme A
CTP	cytidine-5'-triphosphate
CDP	cytidine-5'-diphosphate
CMP	cytidine-5'-monophosphate
cyclic AMP (cAMP)	adenosine-3',5'-monophosphate
DNA	deoxyribonucleic acid
DNAase	deoxyribonuclease
EDTA	ethylenediaminetetra-acetate
EGTA	ethyleneglycolbis(amino-ethylether) tetra-acetate
FAD	flavin-adenine dinucleotide
FMN	flavin mononucleotide
GSH	reduced glutathione
GSSG	oxidized glutathione
GTP	guanine-5'-triphosphate
GDP	guanine-5'-diphosphate
GMP	guanine-5'-monophosphate
IgG	immunoglobulin G
NAD	nicotinamide-adenine dinucleotide
NAD$^+$	nicotinamide-adenine dinucleotide (oxidized form)

NADH	nicotinamide-adenine dinucleotide (reduced form)
NADP	nicotinamide-adenine dinucleotide phosphate
$NADP^+$	nicotinamide-adenine dinucleotide phosphate (oxidized form)
NADPH	nicotinamide-adenine dinucleotide phosphate (reduced form)
P_i	orthophosphate
PP_i	pyrophosphate
RNA	ribonucleic acid
mRNA	messenger ribonucleic acid
rRNA	ribosomal ribonucleic acid
tRNA	transfer ribonucleic acid
RNAase	ribonuclease
SDS	sodium dodecyl sulphate
UTP	uridine-5'-triphosphate
UDP	uridine-5'-diphosphate
UMP	uridine-5'-monophosphate

Phosphates

Phosphates are extremely important compounds in metabolism and a brief explanation of their chemistry is necessary to understand the shorthand forms used to describe inorganic and organic phosphates.

Orthophosphates

IONIZATION

Most phosphates are derivatives of *orthophosphoric acid*, H_3PO_4. This is a tribasic acid and ionizes as follows:

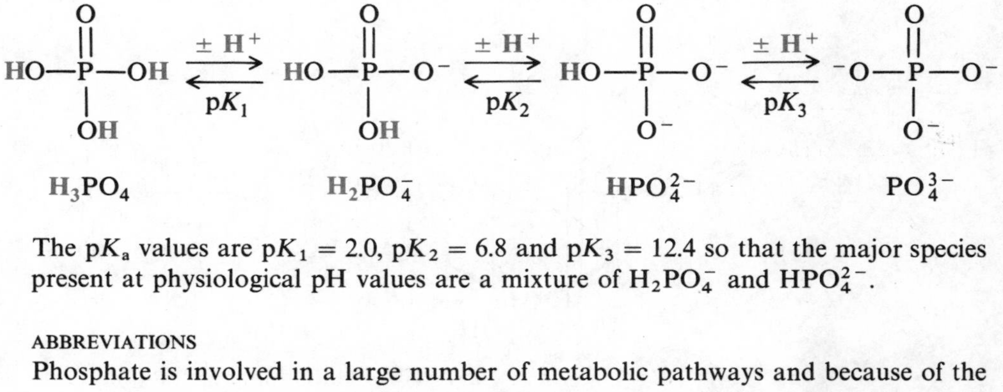

The pK_a values are $pK_1 = 2.0$, $pK_2 = 6.8$ and $pK_3 = 12.4$ so that the major species present at physiological pH values are a mixture of $H_2PO_4^-$ and HPO_4^{2-}.

ABBREVIATIONS

Phosphate is involved in a large number of metabolic pathways and because of the problem of depicting a mixture of ionized forms, abbreviations are used to show the presence of phosphate:

P_i This is the usual abbreviation for inorganic phosphate as the orthophosphate

—Ⓟ   This symbol is used for the presence of phosphate groups on organic molecules

Pyrophosphate

STRUCTURE

The other common form of phosphate is derived from *pyrophosphoric acid*, $H_4P_2O_7$. This is a quadribasic acid and chemically is an acid anhydride formed by the combination of two molecules of orthophosphoric acid with the loss of water.

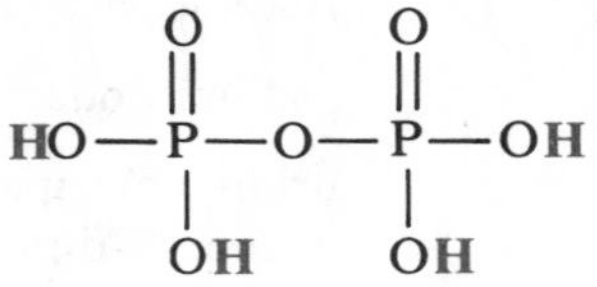

$$HO-\underset{\underset{\displaystyle OH}{|}}{\overset{\overset{\displaystyle O}{||}}{P}}-O-\underset{\underset{\displaystyle OH}{|}}{\overset{\overset{\displaystyle O}{||}}{P}}-OH$$

ABBREVIATIONS

As with orthophosphates, the ionization makes it difficult to write the structure of pyrophosphate and so the following abbreviations are used:

PP_i Inorganic pyrophosphate

—Ⓟ—Ⓟ   Pyrophosphate group in an organic molecule

Adenosine triphosphate (ATP)

A POLYPHOSPHORIC ACID

More than two molecules of orthophosphoric acid may combine with the loss of water to make an extended chain of a *polyphosphoric acid* and an example of this is adenosine triphosphate (ATP):

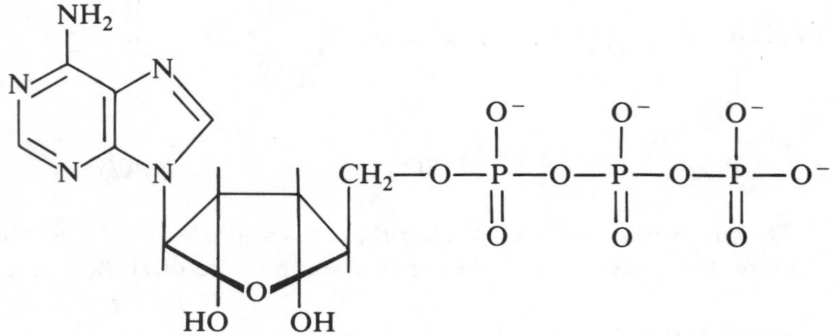

adenosine triphosphate

THE HYDROLYSIS OF ATP

Many biochemical reactions involve the hydrolysis of ATP to ADP and this should strictly be written as:

$$ATP^{4-} + H_2O \longrightarrow ADP^{3-} + HPO_4^{2-} + H^+$$

In practice the charge on the phosphate groups, the molecule of water and the proton are usually omitted and the equation is nearly always written as:

$$ATP \longrightarrow ADP + P_i$$

This simplified equation is convenient but not strictly accurate and it should always be borne in mind that when ATP is hydrolysed water is involved and there is a change in the charge carried by the molecule as well as a loss of a proton.

Index